MORRIS AUTOMATED INFORMATION NETWORK

0 1029 0106383 8

W9-BUI-201

WITHDRAWN

621.
38933
WEE

Weems, David B.

Great sound stereo
speaker manual--with
projects.

| DATE | | |
|---|---|---|
| | | |
| | | |
| | | |
| | | |
| | | |
| | | |
| | | |
| | | |
| | | |
| | | |

Parsippany-Troy Hills Public Library
Lake Hiawatha Branch
Nokomis Ave. & Hiawatha Blvd.
Lake Hiawatha, N.J. 07034

JUN 09 1992

BAKER & TAYLOR BOOKS

# Great Sound Stereo Speaker Manual

## with Projects

David B. Weems

TAB BOOKS
Blue Ridge Summit, PA

FIRST EDITION
THIRD PRINTING

© 1990 by **TAB Books**.
TAB Books is a division of McGraw-Hill, Inc.

Printed in the United States of America. All rights reserved. The publisher takes no responsibility for the use of any of the materials or methods described in this book, nor for the products thereof.

**Library of Congress Cataloging-in-Publication Data**

Weems, David B.
   Great sound stereo speaker manual—with projects / by David B.
Weems.
     p.   cm.
   ISBN 0-8306-9274-6          ISBN 0-8306-3274-3 (pbk.)
   1. Loudspeakers—Design and construction—Amateurs' manuals.
  2. Stereophonic sound systems—Design and construction—Amateurs'
manuals.  I. Title.
TK9968.W428  1990
621.389'33—dc20                        89-29142
                                              CIP

TAB Books offers software for sale. For information and a catalog, please contact TAB Software Department, Blue Ridge Summit, PA 17294-0850.

# About the Author

$D$avid Weems graduated from UCLA in 1949, receiving a B.S. with honors. He worked on research projects in soil physics there for the Engineering Research and Development Laboratory and for the College of Agriculture. While at UCLA, he was introduced to the world of high-fidelity sound by workers in an acoustics lab next door and by a musicologist friend.

In the early 1950s, Weems began working with amplifiers and loudspeakers, trying to improve existing designs. He decided to concentrate on speakers after several experiments that showed promise in that area. His first audio publication in 1954 stressed the advantages of high speaker compliance and showed how to modify the stiff-coned speakers of the era so they could give good bass response in a small box. Then he designed a low-cost speaker enclosure that was featured by an electronics magazine at 1956 audio shows. Following those early publications, Weems has written articles on audio topics for every major electronics magazine in the U.S. Some of his articles have been reprinted abroad — in Australia, Holland, Italy, and India.

After leaving UCLA, Weems taught science in California and then physics and chemistry in Missouri. Since 1968, he has been a free-lance writer. His credits include the training manuals in basic science for the Neosho Water & Wastewater Technical School, a worldwide technical school based in Neosho, Missouri. Dave and Charys Weems have been avid bicyclists since they met at UCLA. They and their daughter, Helen, live in southwest Missouri.

**CREDITS**
Acquiring Editor: Roland S. Phelps
Editorial Team: Lisa A. Doyle, Editor
                    Laura L. Crist
                    Julie A. Ritter
                    Elizabeth J. Akers
Book Design: Jaclyn J. Boone
Production Manager: Katherine Brown
Production: Digitype, Inc.
Illustrator: Robert M. Cox
Indexer: Joann Woy
Cover Design: C. Douglas Robson
                    Lori M. Schlosser
                    Carol A. Sawyer
                    Richard Holberg, Holberg Design

# Contents

# Projects

# Acknowledgments

$S$ome projects in this book were originally published in *Hands-On Electronics* and *Speaker Builder* magazines. I would like to thank Julian Martin, editor of *Popular Electronics* (now combined with *Hands-On Electronics*), for permission to use Project 10 and the information on the impedance equalizer and resonant-peak filter (copyright 1986). I also thank Edward T. Dell, publisher of *Speaker Builder*, for permission to use Projects 4 (copyright 1988), 6 (copyright 1985), 7 (copyright 1987), and the information on notch filters (copyright 1986). (*Speaker Builder* magazine, P.O. Box 494, Peterborough, NH 03458. $20/yr., Audio Amateur Publications.)

I also want to thank D. B. Keele, Jr. for sharing his calculator method of vented-box design, Ralph Gonzalez for some specific suggestions on notch filter design, and Hal Finnell for work on the computer program in Appendix B.

This book, like all such books, owes much to numerous other people. Some of them from whom ideas were borrowed are mentioned throughout the book.

# Introduction

*T*his book is designed to be useful to the passive consumer as well as to the audio activist. Surely the goal of each is better stereo sound.

One aim of the book is to stress those decisions you make in building or using speakers that make a noticeable difference in performance. At the same time, I try to note that there is more than one answer for most audio questions. Considering that, doing things the easy way might not be one of the greater evils of the world. Where possible, the path taken is the simple one.

The first few chapters furnish background and advice on using speakers. The middle portion of the book goes into designing and building your own speakers. Appendix A lists several mail-order houses, and Appendix B is a computer program for those who want to explore speaker design via a PC, which includes driver parameters, vented-box design, and closed-box design. Finally, a complete glossary is provided to explain terms and materials to the uninitiated.

Several years ago, some experts were writing that the day of the amateur speaker builder was over. Since then there has been a revival, both in companies who supply speakers and in the much greater availability of related components, as more and more audiophiles take up the hobby. So much for soothsayers.

Speaker design and construction is a hobby that combines a variety of skills. Maybe that's one reason it is growing. Welcome aboard.

*Chapter* **1**

# Current Trends

$S$uppose Rip Van Winkle were a loudspeaker engineer who went to sleep twenty years ago. He would awake now to some surprising changes in speaker technology. Some of the arguments he had with other engineers would no longer be of interest. They would long have been settled or made irrelevant by advances in our knowledge of how speaker systems work. But the kind of change that might be the hardest for Rip to accept is that some of the cut-and-dried "facts" that he knew have been brought into question.

One of the purposes of this book is to examine the recent changes in stereo speaker development and to sift out the significant ones.

## TWEAKERS vs. THE ESTABLISHMENT

All through the history of high fidelity and stereo sound, a small group of fanatics have pestered the industry into progress. At around the time of World War II, the major radio networks "proved" that the average listener preferred sound with a limited frequency range. The hi-fi nuts of that day didn't believe it. They kept right on building their wide-range amplifiers, often using war-surplus parts. Then Harry F. Olson conducted a landmark experiment that vindicated the nuts. He found that listeners preferred sound of limited frequency range only when it was distorted.

Some audio cranks appear to go off the deep end at times with statements such as the one going around now that speaker cable is directional. In other words, they suggest that you must hook up a new cable one way and then reverse it to determine in which direction it works better. Some smile at such apparent absurdities, but who knows which of the wild ideas could turn out to have merit? True progress is never cheap; it costs time, labor, money, and sometimes all three. It often consists of refining the old rather than inventing something wildly different. Remembering that, let's take a look at current speaker design.

## MODERN DYNAMIC DRIVERS

If you inspect almost any speaker system, you are likely to find a familiar kind of driver—the *dynamic speaker*. Dynamic drivers are speakers with cones or domes that produce the sound. The cones or domes are connected at one end to voice coils that are controlled by a magnetic field. The whole assembly of a dynamic speaker is supported by a basket of stamped steel or cast alloy. Such speakers have been the basic building blocks of sound systems for more than a half century. Not much of a revolution there.

Looking closer, you see more variety than you would have a few years ago (FIG. 1-1). The traditional material for the cone is paper. It is still used by many manufacturers, although it is often coated with various kinds of *dope*.

During the last generation, some manufacturers have substituted

**1-1**   These drivers show the use of polypropylene, titanium, and other materials for their diaphragms.

several kinds of plastic for paper. They contend that plastic is more consistent in composition, unlike paper pulp that can vary from batch to batch. The most popular material now is probably polypropylene. It is made in various shades from black to gray to translucent. It seems to be a top choice of speaker manufacturers in the United States, Europe, and the Orient. They claim it has less coloration than paper and is stiffer as well. A few speaker designers refuse to concede to its apparent superiority, sticking with doped paper cones.

Bextrene was one of the first paper substitutes and is still in use. Bextrene drivers usually show smooth response curves but have lower sensitivity than most other materials. Polydax, a subsidiary of Audax of France, sells several models with Bextrene cones. Audax makes some speaker cones from a plastic they call TPX. They say it is more rigid than polypropylene and better damped. Audax uses a different suspension material with TPX cones that they call Norsorex. Like TPX, Norsorex is well damped with a low internal Q.

Focal uses a material they call Neoflex in some of their speakers. They say it has a lower density than any other cone material, giving it great sensitivity. Neoflex cones, like most polypropylene cones, appear to have smooth upper frequency roll-offs, which is a desirable characteristic.

Some companies make cones that are a sandwich of two or more materials. Eton, a German company, bonds two layers of Kevlar on either side of a honeycomb structure. They call their product Hexacone. Focal also makes some sandwich cones. They consist of a layer of foam microballs held between two thin layers of Kevlar by a synthetic resin. These sandwich cones have great stiffness and internal damping.

Tweeter cones and domes come in various shapes and materials. Most domes are convex but a few are concave. The dome shape is used for mechanical stability, not because the shape has any special radiation properties.

When manufacturers first replaced hard paper in tweeter diaphragms, one of the first substitutes they chose was Mylar. It is still in use by some companies. One material that has stood the test of time is the doped fabric of many soft dome tweeters. Various companies choose other materials such as fiberglass, polyamide, titanium, and even aluminum.

Voice coils have gone through various stages of evolution. Even before the stereo age, a few companies mastered the art of edge-wound aluminum voice coils. They produced the highest efficiency speakers of the time, with large woofers in huge enclosures.

When compact air suspension speakers came into use with their floppy suspensions, it was necessary to make the voice coil longer than the magnetic gap so the coil could move farther and yet maintain the same amount of coil in the magnetic field. For such woofers, it was back to the simple coil, wound with circular copper wire.

Now several manufacturers have revived the technique of making edge-wound coils with flat wire. The flattened wire can be packed more closely than circular wire, putting more turns in an equal area of magnetic field.

The double voice coil is another feature that has been around for years but is now getting more popular. It was formerly reserved for subwoofers, serving to mix the bass of both channels in a single driver. Now you can find speakers with double voice coil woofers in each channel. Focal, for example, makes a full line of dual voice coil (DVC) speakers, starting with a 5-inch model. The extra coil can be used to bolster the droopy bass response that is typical of small speakers, or it can be wired to alter the damping on the speaker without affecting midrange reproduction.

You can get an estimate of the power-handling ability of a speaker by checking the diameter of the voice coil. Small woofers usually have a 1-inch voice coil. Woofers designed for higher power usually have voice coils with a 2-inch or larger diameter. The mass of tweeter voice coils must be limited to allow the tweeter to respond to high frequency signals, so cone tweeters often have a voice coil no larger than ½ inch in diameter. Dome tweeters have voice coils as large as the dome itself, so a 1-inch dome has a 1-inch voice coil.

One recent development in tweeter technology is the use of Ferrofluid in the magnetic gap. Ferrofluid is a viscous substance with magnetic particles suspended in it. It improves reliability by conducting heat away from the voice coil. Another advantage of Ferrofluid tweeters is that the extra damping controls the output at resonance. A simpler crossover network can often be used with such a tweeter because of the improved damping at resonance.

## Electrostatic and Ribbon Speakers

*Electrostatic speakers* are based on the well-known law that opposite electrical charges attract and like charges repel. Typically, an electrostatic speaker has two fixed electrodes with a thin film diaphragm of metallized plastic between them. The membrane carries a positive charge while the fixed electrodes receive the ac signal voltage.

For example, with a 1000 Hz signal sent to the speaker, the charges on one of the fixed electrodes alternates between positive and negative 1000 times per second. The other electrode alternates at the same rate but is always 180 degrees out of phase with the first one. The membrane between the electrodes is attracted to each electrode when it is changed negatively and then repels as the electrode goes positive. The appeal of the electrostatic speaker is that the large diaphragm is driven over its entire surface instead of from a point at one edge as in the dynamic driver. Another advantage is that the membrane can be light enough to respond to high frequencies and yet large enough in area to handle the full audio range without a crossover network.

As attractive as these qualities might be, electrostatic speakers have limitations. First, the speaker's electrical behavior is that of a capacitor. Some amplifiers balk at that kind of a load. The full-range electrostatic speakers are somewhat less than rugged. A single occurrence of overloading an electrostatic can knock a hole in the diaphragm that could later allow arcing at a lower-than-the-original overload point. Most electrostatics *beam* (concentrate) the treble, although some have curved radiating surfaces that mitigate the problem. Lastly, but of importance to many listeners, full-range electrostatics are expensive.

Carver, Apogee, and Magneplanar offer *ribbon speakers*. The ribbon speaker consists of a foil conductor placed in a magnetic field. The Carver "Amazing Loudspeaker," for example, uses a ribbon transducer for the spectrum from 100 Hz upward. It is difficult to produce a large ribbon speaker, but Carver obtained a ribbon area similar to that of an 8-inch cone driver by running a dual ribbon, about ½ inch wide, almost the full length of the 5½-foot panel. For the bass, Carver uses a battery of 12-inch dynamic woofers on a flat baffle.

Although the ribbon speaker has some characteristics in common with the electrostatic, its makers claim some other advantages. The ribbon speaker offers a resistive load instead of a capacitive load to the amplifier. There is no problem of arcing. Ribbons have great clarity but, unless they are very large, have weak bass.

Ribbon and electrostatic speakers are dipoles, radiating both to the front and rear of the baffle. This poses a problem in small rooms, because the speakers must be placed a few feet or more from the wall behind them for best performance. Their sound characteristic is that of spaciousness, but often without a precise stereo image. And if their room position is wrong, the image effect is even more confused.

When you read the advertising copy of any of the drivers described previously, it seems that each has a unique advantage that destroys the competition. It isn't that way. One speaker maker used to say that he would produce one model only, the best one he could make. Even if it were perfect, which it was not, it would probably have been too expensive for some purposes. Anyone who has experimented with a lot of different kinds of speakers has seen cases where a "cheap" speaker sounded better than an expensive model.

Good drivers are only a small part of the whole speaker system. Their sound can be compromised by poor enclosures, bad crossovers, or even careless room arrangement.

## CHANGES IN BOX DESIGN

If our Rip Van Winkle had awakened every ten years and visited an audio show, he would have seen a constantly changing box shape. In the 1960s, up to 75 percent of the cabinets had greater width than height (FIG. 1-2). Most of those were bookshelf speakers, designed to fit on a shelf. About 5 percent were square. In the 1970s, 2 or 3 percent were square and about 50 percent were wider than their height. By his visit in to the 1980s, old Rip would have found a swing

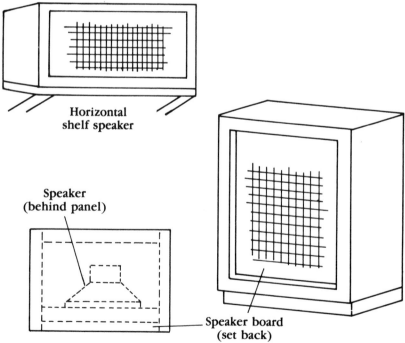

Horizontal
shelf speaker

Speaker
(behind panel)

Speaker board
(set back)

**1-2**  Some outdated methods of building speaker enclosures.

Drivers in vertical line

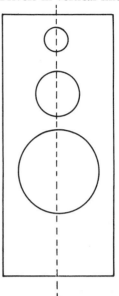

**1-3**  Vertically aligned drivers produce better stereo imaging.

to vertical enclosures, (FIG. 1-3). In one count of 462 models, only 14 had greater width than height. Those few horizontal models almost invariably had special design features that made the greater width manditory. Are these changes passing fads? Or is there a good reason for the preponderance of vertical enclosures?

If there was a single reason for this change of shape, it would likely be a growing concern about phase distortion. There is considerable argument about the desirability of placing the woofer in the same vertical plane as the tweeter, but almost no argument exists about the bad effects of horizontal displacement. Lateral displacement of drivers is easily noticeable because the path length from each driver changes with listener position. Vertically aligned drivers usually have better imaging, as in (FIG. 1-3).

Modern enclosures are not only taller than wide, but some are shaped like columns (FIG. 1-4). They have the advantage of putting the drivers at ear level without the need for a stand. Even so, most speaker boxes are just that, boxes. Most have six walls, the opposite walls parallel. But not all. Some manufacturers go to the expense of truncated pyramids or even pentagonal boxes to depress standing waves in the box. And at least one manufacturer uses an egg shape.

Old magazines show many photographs of expensive built-in

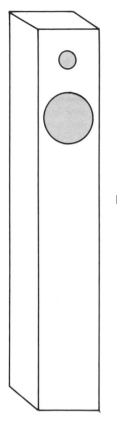

**1-4**    Tall columns raise the drivers to ear level without a stand, and narrow speakers usually have good imaging.

stereo systems. Such systems went directly from the drawing board to installation with no chance to test the effects of speaker location. A big advantage of movable speakers is that you can place them where they sound the best. This is also an advantage for small speakers. Every room, no matter what size, has more possible locations for a small speaker than a large one.

Where plywood was once the preferred box material, particle board has emerged as the clear choice of most builders. All but the most expensive plywood has flaws, such as voids, that affect its rigidity. Particle board appears to be more "dead" to sound, but composite panels made of dissimilar materials, such as particle board and plywood, are also good. Either material can be further deadened by the application of various treatments to the internal walls. Even the kind of finish used on the exterior of a cabinet can affect the kind of midrange sound produced by a speaker system. It is hard to overstate the importance of the enclosure.

Until a generation ago, nearly all speakers were installed from the back side of the speaker board, leaving a cavity in front of the cone (as illustrated back in FIG. 1-2). Today, drivers are almost always installed from the front, and on all but the cheapest speakers, there is some effort to mount the speaker face flush with the speaker board. Some box makers recess the speakers by routing the board; others build up the panel with wood or damping material such as felt.

In addition to smoother driver mounting, considerable care goes into an effort to avoid sharp projections anywhere on the front of the cabinet. Where grille cloth was once set back an inch or so, for the sake of appearance, thick grille frames are now passé. Harry Olson described the bad effects of cabinet diffraction about 40 years ago, but his words went unheeded until recently. He said that box makers should not only avoid projections but that sharp external corners should also be avoided. Many cabinets are now made with rounded edges on the front (FIG. 1-5).

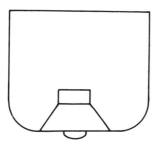

**1-5** This top view shows how some enclosures are made with rounded edges to reduce diffraction.

*Time alignment* is another idea that is given at least some attention. Some builders set the tweeter back; others move the woofer forward in an attempt to equalize the distance from the voice coils to the listener. Still another ploy is to slope the front panel, either by cabinet design (FIG. 1-6) or by installing the box on a stand that tilts.

## CHANGING DESIGN PRIORITIES

At one time, mini speakers were cheap speakers that you used where cost or shape prohibited something better. Now you can pay up to $5000 for a pair. In fact, a pair that sells for about that figure consists of a 6½-inch woofer and a tweeter in each cabinet. And the box holds about a third of a cubic foot of air. Meanwhile, at the local discount store, you can find a three-way system with a 15-inch woofer for $150. What's going on here?

Most speaker designers once put a wide frequency range as their first goal. The engineer chose a large woofer, then a tweeter with

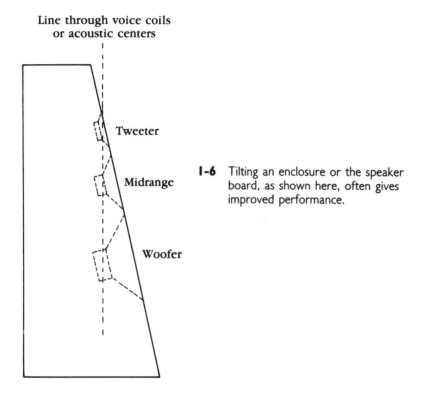

Line through voice coils
or acoustic centers

Tweeter

Midrange

Woofer

**I-6** Tilting an enclosure or the speaker board, as shown here, often gives improved performance.

high highs. Then, as if in afterthought, someone threw in a midrange driver. Presto, a full-range system.

Now more attention is directed at the range that contains most of the music: the midrange. Listeners still pay more for extended bass and treble, but only if it is compatible with good midrange reproduction. What really unlocks the coffers of today's audiophiles is high-definition, precise stereo imaging, and openness. Some small miniatures can knock the average big system dead on those counts.

While there might be a trend to small, and sometimes delicate, systems, there is another trend that pulls speakers in the opposite direction. It is evident in the worn-out phrase "digital ready." What does that mean?

Regardless of the arguments on digital music, pro and con, it is acknowledged that digital recordings require a wider dynamic range than older analog recordings. People who like their music loud might roll up the volume control on the quiet opening passages of an orchestral work to find that the walls are bending at the next crescendo. Such action can blow small speakers. One well-known example of a digital recording that tests the capabilities of speakers is the Telarc recording of Tchaikovsky's 1812 Overture. If you turn the volume up to a comfortable level on the opening, you must accept the consequences of the killer cannon shots farther along—they will tell you if your speakers are "digital ready." Some listeners could play that recording with speakers that housed a pair of 4-inch woofers and be safe. Others, even in a small room, would need much larger drivers. Like many questions, the answer to this one depends very much on personal taste.

To get an idea of how music listening has changed over the last 40 years, consider some tests made by Harry F. Olson about 1950. Olson was director of research at RCA Laboratories then. He said that most people adjusted the volume controls on their radios to reproduce speech at a normal conversational level, or about 70 dB. He set up a demonstration room at RCA that resembled a typical living room in a house or apartment of that time. When visitors came to the "Living Laboratory," he presented programs of music and speech for them. The average sound pressure level (SPL) of his demonstrations was about 80 dB. He said that some listeners objected to the sound level because it was too loud, but no one ever complained that it was too low.

To compare that level with some levels reached today, a rock concert can produce an SPL of about 115 dB. That would be over a thousand times the intensity level that 1950 listeners found "too loud." If you are an addict to rock concert sound levels, you will want to put a high priority on the ruggedness of the speakers you choose.

## CONTEMPORARY CROSSOVER NETWORKS

Crossover network designers once assumed that a speaker's electrical behavior was like that of a resistor. If the label on the driver said "8 ohms," they plugged that number into an equation to find what inductance or capacitance to use in the crossover network. It was well known that the impedance of a speaker varies with frequency, but why worry?

Now crossover designers either consider the actual impedance of a driver at the crossover frequency and beyond or modify the impedance to better control electrical and mechanical behavior. Or the designer might make use of the natural roll-off behavior of a driver in designing a crossover network for it.

One of the surprising things about today's "high-end" speakers is how many are two-way instead of three-way systems. There are several reasons for this. One is the recognition that crossovers can be expensive in design time and in top-quality components. Another is to limit the frequency bands affected by crossover problems. At one time, many speakers consisted of an 8-inch extended range speaker and no tweeter. Such speakers usually had a limited high end, and they often focused treble into a narrow beam. Theoretically, a speaker should be limited to frequencies below that where the diameter of the cone is equal to the wavelength. By that rule, an 8-inch speaker is limited to a maximum frequency of 1700 Hz. Following that precaution, many 8-inch-woofer 2-way systems of a few years ago had a crossover frequency of 1000 Hz. Now the crossover frequency for such a speaker is more likely to be 2500 to 3000 Hz. This change is dictated partly by a desire to move the crossover point above the critical range and partly to limit the range of small tweeters to one they can handle without distortion or failure.

When our friend Rip went to sleep, he could have guessed the price of a speaker by checking the cut-off rate of the crossover network. Low-cost speakers had shallow slope crossovers that cut the stop band of each driver at the rate of 6 dB per octave. More expensive speakers usually had the other kind that was then in wide use — the 12 dB per octave, or *second-order* network. Today, he would be shocked to find some speakers with fourth-order networks of 24 dB per octave for minimum overlap of driver output. But even more puzzling, he would find some very expensive high-end speakers with simple 6-dB-per-octave networks. This variety indicates more independent thinking about crossovers, at least among some current designers.

Even some of the formulas used to calculate the values of the inductors and capacitors for networks are more varied now. For even-order networks with either 12- or 24-dB-per-octave slopes, the exclusive use of the Butterworth filter has given way to the more highly damped Linkwitz-Riley network and others. Butterworth networks produce a flat power response while the L-R networks offer a flat frequency magnitude response.

After the designers finish with a network, the tweakers sometimes take over. They substitute polypropylene capacitors for the cheaper nonpolarized electrolytics and put in oxygen-free wire instead of off-the-shelf hook-up wire. Then they match parts by close tolerances. In general, such efforts improve performance, but sometimes tweakers find that the sound is mysteriously degraded after putting in better components. Chapter 8 explains why and what to do about it.

One change that would surely startle Rip is the variety of exotic speaker cables on the market now. He would have been concerned about wire resistance if he used a cable with small-diameter conductors. Now, consideration is given to other wire characteristics such as inductance and capacitance. In general, high-priced cables have less inductance and greater capacitance than the cheap lamp cord that most audiophiles once used. How does this affect the sound?

The inductance of cheap cable, which is in series with the full speaker system, can make for a slight roll-off in the upper highs. Critical listeners might notice the effect. On the other hand, the capacitive load of some high-priced cables can cause oscillation in some amplifiers, but probably not in those produced today. This appears to be one situation where compatibility is not very predictable.

Old Rip would be even more confused by bi-wiring. Some speakers are wired so that the woofer section and the tweeter section have separate terminals. This allows separate cables to be used for the lows and highs. Users claim that bi-wiring improves performance, particularly in high-frequency detail.

## LISTENING ROOMS

Except for Mr. Klipsch's horns, you rarely find speakers in corners any more. Corners are good positions for maximum bass response, but they are bad in other ways. For similar reasons, most floor speakers no longer stand against a wall, unless furniture arrangement dictates that position. Most speakers come alive then you move them out into the room. And even if the bass response is reduced, its overall quality is better.

You can improve small speakers by installing them on good, solid stands (FIG. 1-7). The worst location for most small speakers is on the floor. Musicians play their instruments at chair level or above, so even if the speaker performed well on the floor, the music would not sound natural to careful listeners. And the chances are that no small speaker can sound its best on the floor. The high frequencies get blocked by other furniture, and reflections off the floor can produce confused sound.

If you are interested in good sound, approach the task of setting

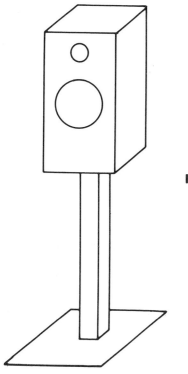

**I-7**  Small speakers should be placed on a well-built stand.

up a speaker system differently than that used by an interior decorator. The decorator would place the furniture for visual balance; you will want to place it for balanced sound. That means the speakers must take priority in room arrangement.

As a listener, you must train yourself to be a more acute listener. Think about what you are hearing. How is it different from what you heard with the speakers in another location? And so on. Chapter 3 explains more on how to get the most from your speakers.

*Chapter* **2**

# How to Buy Speakers

*Y*ou can approach buying any product by emphasizing brand loyalty, following someone's recommendation, or by price. The best way is to analyze your own needs and shop accordingly. To do that, it helps to narrow the field.

## KINDS OF SPEAKERS

This book focuses mainly on home stereo speakers. You have much more choice of models and control over their use than, for example, car speakers. Car speakers have to fit the limited mounting space and sites in your car. With home stereo speakers, you have much more latitude; you can buy a good set of speakers and install them so they produce sound that could be either superb or mediocre.

Your stereo speakers probably consist of a set of dynamic drivers in boxes. The enclosures are likely to be one of two common types, the closed box or the reflex box (FIG. 2-1 and 2-2). Closed boxes are sealed airtight, so the woofer works against a cushion of air in the box. Reflex speakers have ports in the boxes that are tuned to a precise frequency to extend the bass range of the driver. Other expensive speakers use unusual enclosures such as the transmission line or horn.

Many years ago, nearly all high-fidelity speakers had stiff cones and were installed in large floor enclosures with internal volume of 5 cubic feet or more. In the early days of stereo, the need for two

**2-1** Project I: Closed-Box Speakers.

speakers put a premium on space. At the same time, the development of the acoustic suspension speaker, a highly compliant woofer in a compact closed box, fit the bill. By the early 1970s, nearly all home speaker systems were some form of acoustic suspension speaker. Although many of today's speakers are installed in closed boxes, there are differences between the earlier models and current ones.

One way closed-box speakers differ from each other is in the damping on the cone of the woofer. The total damping of the driver's resonance is dependent on the degree of electrical and mechanical damping by the system. An undamped speaker booms loudly at its resonance frequency, producing a bump in the response curve at that point. Damping is stated in the value of the Q of the speaker. The Q value varies inversely with the damping ratio. High damping ratios, or low Qs, require speakers with larger magnets and enclosures than a similar speaker that is underdamped. Be aware that a "large" enclosure, as used here, might not look large. It can be large acoustically while remaining in the compact class. It all depends on the physical and electrical characteristics of the speaker.

At one time, it was generally accepted that a closed box speaker

**2-2**  A ported-box speaker system (Project 5).

should have a Q of about 1. For a given driver, that value offered extended bass range with only moderate peaking at resonance (FIG. 2-3). Some designers aimed for a final Q of 0.7 because that gave the flattest response curve down to cut-off with no peaking. Some critical listeners who value bass purity over bass range prefer overdamped speakers with even lower Qs. At the other extreme, some manufacturers deliberately produce speakers with a Q greater than 1 to give a prominent bass response that, they hope, will sell more speakers. You can often hear the thump of this kind of speaker at department stores. Some rock music fans like this kind of bass.

At one time, bass reflex speakers carried the nickname "boom boxes." One reason was that many speaker designers were not even aware of some of the factors that influenced the performance of a ported system. They knew how to tune the box to a given frequency, but they had to use rules of thumb for the optimum cubic volume in the box or the ideal tuning frequency for a particular driver. The alternative was to build an endless number of enclosures for each driver, test them, and choose the one that performed the best. It's no wonder that most manufacturers had abandoned the reflex speaker by 1970.

**2-3** How Q affects bass response of closed-box speakers with identical resonance frequencies.

In the early 1970s, there were reports in this country of an analysis of reflex speakers done a decade earlier in Australia by Neville Thiele that was based on still earlier work by James F. Novak of Jensen. Following that, Thiele's work was expanded by Richard Small, D. B. Keele, Jr., and others. The outcome of all this was a new life for reflex speakers as designers began to understand how to theoretically match a box to a driver.

Partisans of the ported box argue that the bass loading produced on the cone by the port reduces distortion over a band of bass frequencies. This loading damps the cone and reduces how far the cone must travel to radiate bass. A reflex speaker the same size as a closed-box speaker can have a more extended bass range.

One obvious disadvantage of the reflex system is that the driver must be carefully matched to the enclosure to obtain the desired performance. Another is that the bass cutoff rate is sharper for the reflex than for the closed box so that a closed-box speaker with the same bass cutoff frequency as a reflex can produce usable bass at a lower frequency. Reflex speakers are somewhat vulnerable to infrasonic signals because bass loading is reduced below the port resonance frequency. This allows turntable rumble, or other ultralow frequency signals, to produce cone flutter in the driver. Such movement, which contributes to speaker distortion, can be prevented by a proper infrasonic filter on the amplifier.

The transmission line speaker consists of a driver that is mounted at the end of a long tunnel. The tunnel is stuffed with acoustical damping material to depress the resonances that are normal to an undamped pipe. Some transmission lines make use of port radiation for bass reinforcement. Others are designed for total absorption of the back wave. Transmission line advocates dwell on the quality as much as on the range of their bass. They say the reflex, with its sharp cutoff, is subject to ringing. They also dislike closed boxes, calling them "pressure boxes." Many listeners find that transmission lines have obvious disadvantages for them: the lines are usually large and expensive. The remaining types of home speakers represent a small minority of the total population.

## Miscellaneous Speakers

In recent years, a large market has developed for good-quality car speakers. One reason is the kind of cheap speakers put in cars by the factories. Where original speakers usually have untreated paper cones and magnets of a few ounces, you can buy car speakers with coaxially mounted dome tweeters and large magnets of 20 ounces or more. The main problem is getting speakers to fit the space available in your car.

Musicians need a special kind of speaker: the musical instrument speaker. Such speakers are built to take rough treatment, both physical and electrical. A woofer for musical instrument use has a higher frequency of resonance and a stiffer cone than a high-fidelity speaker. This limits the bass range of such speakers, but the stiff suspensions and special heat-treated voice coils can take the shape transients of instruments such as the bass guitar without failure. Musical instrument speakers have high efficiency. They are often used in batteries that are fed by amplifiers of enormous power output to fill large auditoriums with sound. They are designed to play loud, not to delineate subtle tones.

You could use a musical instrument speaker in your living room if you wanted to, but you would probably be disappointed in its performance. A home stereo speaker used as a musical instrument speaker would be destroyed.

Public-address systems use a wide range of speakers from full-range 8-inch cones to all-weather horns and column speakers. The better PA speakers are designed to have controlled dispersion, particularly in the vertical dimension. Such speakers are designed to give good coverage of large areas without producing feedback in the microphone-amplifier-speaker circuit. Ordinary speakers, such as home stereo speakers, might serve poorly as PA speakers.

## LISTENING ROOM REQUIREMENTS

In absolute terms, a good speaker should sound good anywhere, so the kind of room you have shouldn't matter. However, in practical use, it often does. As an extreme example, if you have a small listening room with no available corners, where would you put a corner horn? Or, suppose you buy two dipole speakers and try to place them in a small room. Dipole speakers must be placed some distance from the rear wall. Add to that the fact that dipole speakers are usually very large, and you can see the problem.

One rule of thumb that works fairly well is to choose a size of speaker that will look right in the room. If a speaker looks out of scale, it probably can't be installed for best performance. Two large cabinets in a small room can occupy so much space it will be difficult to separate them far enough (to realize their stereo potential) and yet avoid placing them too near a reflecting wall. One advantage of small stereo speakers is that you have more latitude in room placement. You can experiment until you find the locations that produce the best stereo illusion.

One family I know made a speaker choice based on the amount of space available at the two ends of a built-in bookcase. The space required a narrow floor cabinet about 27 inches high, so they considered that requirement first. If space or visual effect is very important to you, let it be the ruling reason for your choice of speakers. One of the advantages of designing and building your own speaker system is that you can customize it to fit your needs.

One way to solve the space problem is to install a subwoofer and two satellite speakers instead of a pair of full-range speakers. With a low-frequency crossover, the subwoofer can be hidden or built into existing furniture.

One of the most difficult problems is to choose a speaker that will produce the sound pressure levels you require without spending more money than necessary. It's hard to predict the required loudness because it will be affected by the size of your room, its liveness, and, most of all, by how you listen.

As mentioned in Chapter 1, some listeners think that 80 dB is too loud. Most music lovers would not. Almost everyone who can be called an audiophile has heard a large piano played at full volume and can tolerate that level of sound. Some performers reach levels of about 100 dB with a piano in a room of average size and furnishings. To give a fully natural reproduction of such music would require a speaker system capable of 100 dB or greater output. Note that a level of 90 dB requires 10 times the power needed at 80 dB, and 100 dB requires the power 100 times that of 80 dB.

Few people choose to produce the sound of a rock concert or that of a full-size orchestra at the distance of only a few feet (in their living room). To do either, a speaker system must be capable of producing a sound level of 110 to 115 dB. To radiate this amount of acoustic energy requires enormous volume displacement. *Volume displacement* is the volume "throw" traveled by the driver's piston and is equal to the area of the cone multiplied by the maximum cone deflection. As an example of the kind of speakers required for anything near this kind of performance, the Swan IV speaker system is claimed to produce an SPL of 110 dB throughout the audio spectrum. It employs four 12-inch woofers in two reflex enclosures tuned to 22 Hz. Woofers in closed boxes would have to have greater volume displacement to achieve the same sound level because the Swan woofers are augmented by the ports at the low end of the spectrum where the volume displacement is greatest.

At the other end of the scale, for background music or in a dorm room, mini speakers can fill a need (FIG. 2-4). Moving up in size, a pair of speakers with 6½-inch woofers can give very satisfying sound. Just don't expect to rattle the windows with them. But if you like the sound of a full organ or want to recreate the cannon shots of the 1812 Overture at a level that pushes your insides about, you will need larger woofers.

If you try to use any speaker system beyond its limits, it will suffer dynamic compression and distortion. Regardless of the requirements for high SPLs, the fact remains that many listeners are quite satisfied with the sound of good, small speakers. Those listeners who value transparency, clarity, and a stable stereo image over sheer power will likely be just as happy with a small speaker as a large one.

**2-4**   Mini speakers can provide good stereo listening in small rooms or anywhere that a high sound level isn't needed (Project 2).

## KINDS OF AUDIO EQUIPMENT

The kind of equipment you use with your speakers can make a difference in the kind of speaker you should choose. Here is an example: one listener I know has a fairly low-cost receiver with a power output rated at 35 watts per channel. When used within its power range, this receiver compares favorably with some much more expensive ones in all ways but one. Its one shortcoming is in its infrasonic filter. That filter has a cutoff frequency far above that of two other more expensive receivers against which it was compared in a listening test (FIG. 2-5). With the filters out of the circuit, the three receivers were judged about equal in listening quality. But when the infrasonic filters were switched on, receiver C had an apparent loss of bass response. If you were buying speakers for that receiver, it would be advisable to choose a closed-box speaker rather than a ported system.

With a ported system, you would have to choose between lim-ited bass response or possible speaker overload by turntable rumble

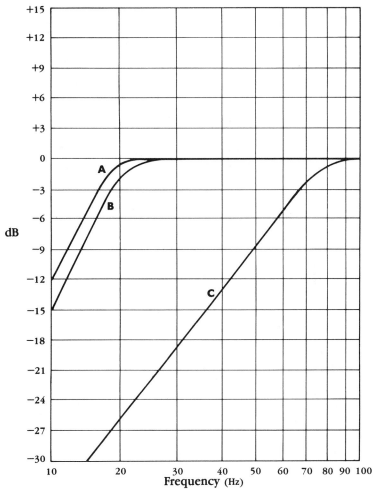

**2-5**   The bass cut response of the infrasonic filter on three receivers.

or other low-frequency garbage. If you are buying speakers to use with a receiver you already own, check the instruction booklet that came with the receiver for a description of the infrasonic filter. If the cutoff frequency is much above 20 Hz, you will probably hear the difference with the filter switched on. If you can't find the book, try listening with the filter on and off. Make sure you use several program sources for such a test; some program material has little to offer under about 50 Hz, so you wouldn't hear any difference either way.

If you own a low-powered amplifier or receiver, check the efficiency rating of the speakers you buy. Efficiency is one characteristic in which speakers vary enormously from brand to brand or even model to model. Manufacturers of commercial speaker systems often give a recommendation for a minimum power rating in the amplifier to be used with their speaker. Unless the speaker is to be used purely for background music, this rating should be observed.

If you are buying a new receiver or amplifier, don't worry about getting one that is too powerful. Chances are that you are more likely to blow speakers with one that has *inadequate* power output. If you try to get a 50-watt output from the speakers via a 15-watt receiver, you can drive the receiver's amplifiers into the clipping range. When that happens, the pile-up of upper-frequency harmonics sends much more power to the tweeters than would exist in normal music, blowing them. Remember this: low-powered amplifiers blow tweeters, high-powered ones rattle woofers.

As a rough estimate, if you have a typical listening room and speakers and you are a typical listener, you might need 100 watts of amplifier power to drive your speakers for realistic sound reproduction. That is total. You might think that a 50-watt-per-channel amplifier can provide that kind of power, but it doesn't always work that way. In some music, one channel could carry more than half the power, so it's a good idea to make each channel capable of handling 75 percent of the total. That would suggest a 75-watt-per-channel amplifier. But suppose you like your music 10 dB above that of the average listener. You would need ten times that much power. If you double the power of your amplifier, you add 3 dB to the sound level, just enough to notice.

The kind of program sources you use will affect the amount of amplifier power you will need. Compact discs have more dynamic range than LPs, for example, and you will need more power headroom. An infrasonic filter shouldn't be necessary with CDs because there will be no rumble on CDs unless they are copies of old recordings.

Even with CDs, you can get satisfying sound from small speakers if you use a bit of caution. A small speaker can offer flat response to a low frequency if the power output at that frequency is limited to a moderate sound level. That is why personal taste in music loudness means so much in choosing a satisfactory speaker system.

## WHERE TO BUY

You can find speakers in many kinds of stores. If you are reading this book, you obviously aren't willing to settle for mediocre sound. That pretty much eliminates such sources as department stores and discount stores. To see what you are missing, stroll by the electronics department of one of these emporiums and rap the back of a speaker cabinet. In most cases it will rattle. At best it will probably sound like a hollow drum.

If you go to a local audio dealer, you will find better merchandise, but such a place is no guarantee of a good buy. Unfortunately, sales people are often more interested in meeting a quota than seeing that the customer makes the best choice. That quota could also include a demand from management that they push a certain piece of equipment regardless of its quality. They sometimes resort to tricks to sway you toward the speaker they want you to buy. One way is to make sure the speaker they are pushing is driven to a higher sound level than the ones to which they are comparing it. Another is to keep going back to their "pet" speaker, particularly for a dramatic piece of music. Sales people do such things as boost the bass or push in a loudness control when that speaker comes on. I once watched a salesman demonstrate CD players in an obviously unfair way. While demonstrating player A, he kept touting player B as more reliable. Then he rolled down the volume control and removed the disc from player A and put it in player B. He raised the volume control to a higher level than with A. The crowd in the showroom immediately agreed that player B sounded better than player A.

Another way stores sometimes try to unload questionable merchandise is to make a package deal that includes a complete component sound system. The price is often attractive, but it usually isn't if you check the quality of the speakers included in the deal.

There are advantages and disadvantages in buying from mail-order dealers. You can't see the merchandise by mail order, but most dealers guarantee their products. Some offer counseling by telephone, if you pay for the call. At least one, Crutchfield Corporation, even allows you to call for technical help on an 800 number if you have bought speakers from them.

The range of prices in commercial speaker systems is rather wide. At the bottom end of the pile, you can buy any number of cheap stereo pairs for $100 or less. They will have thin-walled cabinets and shiny foil domes in the centers of the speaker cones. At the upper end, there are systems that sell for many thousands of dollars. Most music listeners don't even know that those speakers exist.

You surely want to choose a speaker system that offers good value. One way to get more sound for your money is to build your own speaker system. Starting with Chapter 4, this book gives procedures for design and construction of speaker systems.

## DON'T GO TO EXTREMES

It pays to reject the fallacy that there is only one way to produce a top-flight speaker system. In some ads you are asked to believe that certain drivers are indispensible for good sound. Another ad tries to convince you that only a certain kind of crossover network can give natural stereo reproduction. Still another company suggests that its special driver arrangement in the enclosure makes it superior.

The fact is that every speaker system is a compromise between conflicting requirements. A successful speaker designer is one who by careful planning, or even luck, makes the most practical choices for a given size and cost. However you get your speakers, by building or buying, their performance depends on how you set them up and use them.

*Chapter* **3**

# How to Get
# Great Stereo Sound

*A* good stereo system is like a magician. A great magician provides the illusion of performing miracles. He can pull a rabbit out of a hat or a songbird out of thin air. A stereo system can give the illusion that the musicians are there in the room with you. If you close your eyes, you can see them.

## LISTENING ROOM

There are many links in the audio chain. Your room is one of them. The science of acoustics has always been considered more art than science by some practitioners. For proof, they cite several well-known, carefully planned concert halls that were almost unusable when completed. Nevertheless, you should be aware of a few basic rules.

Unless you are shopping for a place to live, there isn't much you can do about room dimensions. But be warned that you will have problems if the room is a cube. When all three dimensions are the same, the standing wave build-up of the fundamental resonance and its multiples occur at the same frequencies for each dimension. The next worst case is a room with two dimensions that are equal. Following that, a room with one dimension that is an exact multiple of another is undesirable.

The ideal room would have dimension ratios that produce evenly spaced resonances within one octave. By spacing the fundamental resonances within one octave, all multiples of those resonances will be spaced throughout the entire audible range. This kind of spacing is rare with large rooms because most rooms have 8-foot ceilings. For a given height, only one room size can have an optimum dimension ratio. One suggested ratio is 0.79 : 1 : 1.26. Hence, for a room with an 8-foot ceiling, the other dimensions would be 10×12⅔ feet. Obviously very few rooms meet this test.

If you have a houseful of rooms to choose from, don't automatically set up your stereo system in the living room. First consider which room offers the quietest acoustical environment. I once lived in a house that had a kitchen just off of the living room. With the stereo system in the living room, you could hear the cycling of the refrigerator. My audiophile friends took to unplugging the refrigerator every time they visited, sometimes forgetting to plug it in again. Air conditioners and forced-air furnaces are also likely sound polluters.

Except for the presence of furnaces, basements are a good choice. You are farther from the noises of the street, and the solid walls and floor make for better bass performance.

If you cannot control the shape or size of your room, you can alter its acoustical environment. Unless you are spending all of your money on sound equipment, your room surely won't be bare. A bare room is a bad room, like a poorly designed boom box. In acoustical terms, a bare room's sound decay time would be too long. If you have ever been in a bare concrete room when a telephone rang, you can understand the problem. Basement walls particularly need to be treated.

It might be good to have the reverberation time the same for all frequencies, but that is almost impossible. Upper highs are always absorbed by the surroundings more easily than low or midrange sound. Unless the relative humidity of the air is like a desert, the frequencies above about 3000 Hz are absorbed before they have time to reverberate. In a practical room the decay time can vary for different frequencies. For example, a room with an overly long decay time of 3 or 4 seconds at 1200 Hz might have a decay time of only 1 second at 1500 Hz. Even concert halls have decay times that decrease with frequencies over 2000 or 3000 Hz. Small rooms with overstuffed furniture, thick carpeting, and draperies can be even more dead to high frequencies.

In setting up a room for ideal listening conditions, try to observe symmetry. If one speaker faces a section of the room that is more heavily draped than the other, your stereo sound will suffer. It is easy to rotate the balance control until the output from the speakers appears to be equal, but the decay time for the two channels will be different. Ideally, the reverberation energy should appear to come to the listener equally from all directions. Such perfection is rarely attainable, but a gross difference will degrade the stereo illusion.

## ROOM POSITION

There is probably nothing, except the quality of the enclosure, that has as much effect on the sound of a speaker as its room position. I recently saw an example of that. An audiophile was testing two stereo speakers to see if they were working properly, so he placed the two floor models side by side. They stood near one end of a 13-×-20-foot room. He adjusted the tweeter controls until they were well balanced, then put on a Linda Ronstadt record. When he switched to mono and turned the balance control back and forth, there was a noticeable difference in Ronstadt's voice from the two speakers. It appeared to be more colored from the left speaker.

The owner switched the crossover networks, but that didn't help. He then painted the gray paper cone of the left woofer with a thin coating of silicone rubber, a treatment that sometimes improves cheap paper cone speakers. No difference. At that failure, he replaced the coated woofer with a new one. The coloration remained.

As a last experiment, he took the speakers into the next room and rearranged the damping material in the left speaker. No change. Then, next day, he took them back to the original room and listened again. There it was, the same coloration in Linda Ronstadt's voice from the left speaker. During the alterations he had scratched the left speaker, so it was easy to identify. And it was now to the right. The speaker that stood where the left speaker had been was the right channel speaker. He reversed the cabinets without changing the wiring, and once again, the problem skipped from one box to the other. During all the tests, the speakers had been so close together they almost touched, so he had not suspected the real culprit: room position. It would have been less surprising if the problem had been in the low bass.

In the ideal listening arrangement, your speakers should be placed at the corners of a triangle made by the speakers and the listening area (FIG. 3-1). Note that, if possible, the speakers in the two channels should be equidistant from the rear wall ($D_1 = D_2$) and the side walls ($D_3 = D_4$). It usually helps to "toe-in" the speakers. The angle that one speaker makes with a line parallel to the rear wall should be equaled in the other speaker, $\theta = \theta'$ in FIG. 3-1. One way to

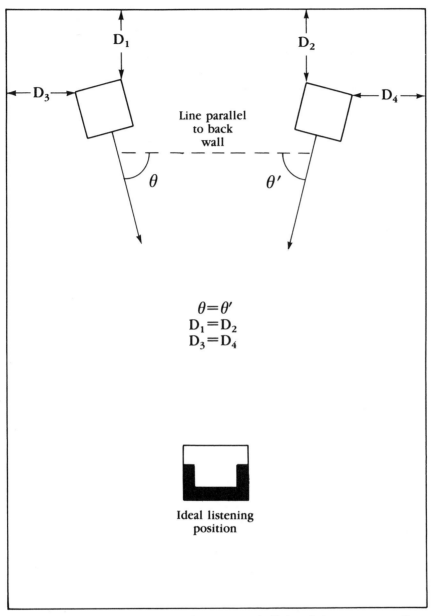

$$\theta = \theta'$$
$$D_1 = D_2$$
$$D_3 = D_4$$

Ideal listening
position

**3-1** If possible, arrange your speakers symmetrically in relation to room walls.

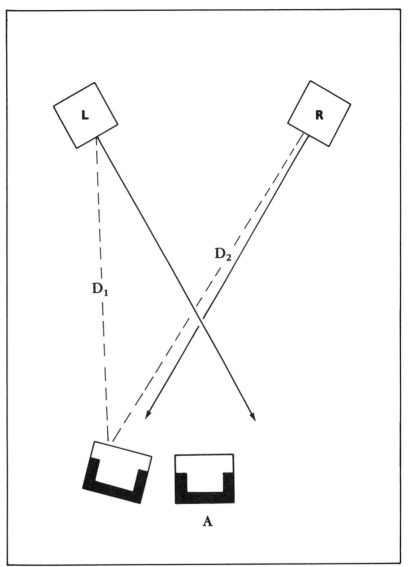

**3-2** Greater toe-in can broaden the stereo image. Listener at B is closer to the left speaker but on the axis of the right one.

widen the area of stereo perception (to include more than one listener) is to *increase* the toe-in (FIG. 3-2). The listener at position B in the figure is closer to the left speaker than the right one, but the normal pull of the stereo image to the left might be nullified by the fact that B is on the axis of the right speaker. Achieving such an effect requires considerable experimentation, but if several people plan to listen, it is worth trying.

Even a slight variation in speaker angle can alter the stereo image. No speaker, no matter how expensive, has perfectly uniform dispersion at all frequencies. Because of uneven dispersion, a slight difference in angle can affect the perceived frequency response of the speaker at the listening area. In all cases, you must experiment to find the optimum arrangement for your situation.

## THE IMPORTANCE OF SYMMETRY

To see why symmetry is important in setting up speakers, consider the precision to which a sound can be located and how it is affected by frequency response. If the speakers in the two channels have identical frequency response, a highly localized sound will be heard as a point in space (FIG. 3-3). You can point to it as surely as if you could see the instrument. Tests have indicated that if the speaker in one channel differs from that of the other channel by only 1 dB, the apparent source is displaced by about 2 degrees. At a listening distance of 10 feet, that can move the apparent source by a few inches. For a pair of speakers with unequal frequency response, the source could move about as the frequency of the music changes. Except for rock concerts, musicians don't normally wander over the stage as they play. Ultimately, an unstable stereo image reduces the stereo experience.

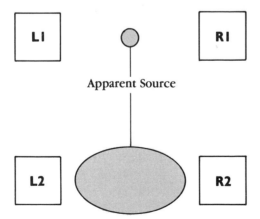

**3-3**   The first set of speakers, L1 and R1, produces a more precise stereo image than L2 and R2 because they are more equally balanced in frequency response.

When the stereo age arrived, many listeners who already owned a large mono speaker simply bought a smaller second speaker to go stereo on the cheap. It seemed to work for demonstration records of ping pong games, but gradually it dawned on those listeners that they were missing some subtle aspects of stereo. One advantage of stereo over mono sound is that you can pick out the sound of weak instruments from the massed voices of an orchestra by their location. Unless the stereo reproduction is precise, lost instruments stay lost, and strong instruments can grow in width (FIG. 3-3).

When you are setting up your speakers, follow the path of symmetry in every way you can. This rule applies to drivers, enclosures, and wiring as well as room position.

## SPEAKER STANDS

If you choose small speakers, you will need stands for them. Even speakers of intermediate size sound more open if they are raised slightly and tipped back a few degrees from the vertical. You can estimate the optimum height and angle by placing some large books on the floor to raise and tilt the speakers. The plan shown in FIG. 3-4 works well with compact speaker enclosures of about 1½ to 2 or more cubic feet.

There is a trend now to make stands adhere to the floor by placing spikes under them. The spikes penetrate the carpet and stick into the wood flooring. With the mass of the speakers plus that of the stands concentrated on the points of the spikes, the structure resists movement. Theoretically, a movable speaker loses bass response by shifting slightly as the woofer pumps out bass. For solidity, some hollow stands are filled with sand. As a cheap substitute, you can build up stands with concrete blocks or bricks.

## SPEAKER CABLES AND CONNECTIONS

If your budget is tight, you might want to choose a low-priced speaker cable. Ordinary 18-gauge lamp cord works fairly well. Don't use old wire that has become oxidized. You can buy "oxygen-free" cord at very reasonable prices, and it should be good enough for starters. Don't forget to buy enough cable to use equal lengths in each channel regardless of the distance from your receiver to each speaker. Symmetry again.

One problem with fancy cable is that it seems to be amplifier sensitive. A bigger problem, for most listeners, is the price. Before squandering too much of your sound budget on special cables, pay

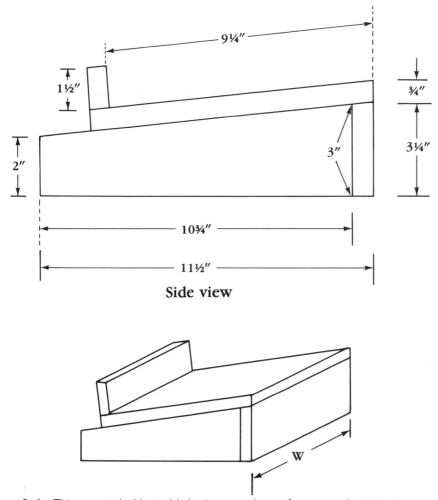

9¼″

1½″

¾″

3″

3¼″

2″

10¾″

11½″

**Side view**

W

**3-4**  This easy-to-build stand helps improve the performance of middle-size speakers by raising and tilting them.

some attention to the kind of connections in your system. Many commercial speakers use 5-way binding posts. They will accept banana plugs, spade lugs, straight wire, or wire made into a J. Banana plugs might work well as speaker connectors for a while, but after being pushed in and out several times they begin to loosen. Spade lugs are better. Some connectors are gold-plated for protection against corrosion. An investment in a good set of terminals is probably a wiser choice than exotic cable with ordinary terminals. In fact, the most likely spot for trouble in the wiring system is at the terminals of the receiver or speaker. Ideally, you should wear gloves when handling the matching parts so that no oil or sweat from your hands contaminates the copper or other metal in the connectors. Dirty or corroded terminals can significantly degrade performance.

One type of popular speaker terminal uses spring-loaded contacts. These seem to be adequate when new, but screw-on terminals are better.

## TO SOLDER OR NOT

Some speaker builders advise against using solder in speaker connections because it is not a good conductor. They prefer to crimp spade lugs to the cable and to use push-on connectors at the driver terminals. Soldering has been the traditional way to make electrical connections and for good reason. Solder seals the joint so that air can't get into it and cause corrosion. Some solderers add a final step. They hold a waxy crayon to the hot solder so that it flows over the surface, making still another layer of sealant. (Solder alone should never be used to hold two conducting wires together.)

To solder properly, carefully clean the matching metal surfaces, make a tight physical connection, and then solder. Don't transport the solder to the work on the soldering tool. Instead, hold a freshly tinned tip under the joint to heat it. Then apply the solder from above the joint so it heats through before the solder flows. If done properly, the joint will be smooth and shiny.

Choose a good quality solder designed for electrical work. It should have a rosin core and a tin/lead content of about 60/40. Don't use a soldering tip that has ever been used with acid core solder. Avoid using large amounts of solder. If you solder the terminals of small tweeters, be careful not to overheat them. Push-on connectors are useful there.

## BI-WIRING: One Way to Tweak

Bi-wiring is a possibility only if you have a speaker system that was designed for it. If you build your own speakers, you can add this feature.

One theory behind bi-wiring is that you can choose the best kind of cable for bass performance in the woofer circuit and the best kind for treble in the tweeter circuit. For low-cost bi-wiring, use the kind of cable used in house wiring for the bass and ordinary speaker cable for the treble. Romex cable labeled 12/2 or the even heavier 10/2 gauge is appropriate for the high-current demands of 4-ohm woofers.

Some experimenters recommend running even more wires from the amplifier to the speakers. They say the ground point of each branch of the crossover network should have a separate return to the amplifier. This makes for a lot of wiring. The ideal way to hide such wiring is to run it through the walls or under the floor.

Is bi-wiring worthwhile? There are several schools of thought on this. One group says you can hear the difference immediately and that it is good. Another says that the value of bi-wiring depends on the complexity of the crossover in the speaker system. Simple crossovers, they say, don't need bi-wiring. James Boyk, a lecturer in Electrical Engineering at Cal Tech and a pianist, reported in *Stereophile* that he tested bi-wiring and even quad-wiring with one speaker system and concluded that the stock speaker was better than the multi-wired speakers. He suggested that the crossover branches in a carefully designed speaker system interact in a beneficial way. If so, bi-wiring would upset the finely tuned design.

## REWIRING: Another Tweak

Most commercial crossover networks use nonpolarized electrolytic (NP) capacitors. The designers of high-end speakers choose more expensive kinds such as Mylar or even polypropylene. Speakers of intermediate price might have nonpolarized electrolytics in the woofer circuit and Mylar for the signal path to the tweeter. This is the practical approach because large values of the better grade of capacitors are not only costly but occupy considerable space. It is generally agreed upon that good, nonpolarized electrolytics are satisfactory for low-frequency duty.

Tweakers sometimes rewire their crossover networks, replacing the nonpolarized electrolytics with Mylar or polypropylene types of the same value. They find this kind of replacement nearly always changes the sound. Careful listening might reveal that the change isn't an improvement.

There are two reasons why better capacitors might not make for better sound. A Mylar capacitor that is marked as a $4\mu F$ capacitor might not be equal to an NP electrolytic labeled with the same value. I have found most electrolytics to be lower in value than Mylars with the same rating. Others who have checked values have sometimes agreed on that; sometimes they have argued that Mylars are lower. Either way, there can be a difference. Any change in capacitance could make a difference in frequency response, but the difference probably won't be significant.

The more important change is likely to be related to a difference in internal resistance of the various capacitors. Nonpolarized electrolytics have greater internal resistance than equivalent Mylars. If you switch capacitors in a speaker system that has fixed resistors to balance the tweeter output to the woofer, the effect will be the same as if you altered the value of the resistors. This will upset the woofer/tweeter balance. If the system was properly balanced with the electrolytic capacitor, it will be too "bright" with the Mylar. If the speaker has an L-pad in the tweeter circuit, there is no problem. Adjust the pad to correct the imbalance.

If you want to upgrade the capacitors in your crossover networks but don't want to make major alterations, there is an easier way. You can bypass the nonpolarized electrolytics with small-value Mylar or polypropylene capacitors. To bypass, you simply wire the added capacitor in parallel with the original. Note that this will change the net capacitance, so you should use a small-value capacitor for bypassing, from 0.1 to 1 $\mu$F, depending on the value of the original capacitor.

Some crossover networks have cheap inductors that are wound from small-diameter wire. These add resistance to the circuit. It probably isn't worthwhile to start changing them; you might just as easily install a new crossover network. And remember there is always the chance that the designer made use of the resistance of the coil to set the speaker damping on the woofer.

## SPEAKER BALANCE: Everyone's Tweak

Most people either leave the left channel/right channel balance control in the center position or set it casually and forget it. *If* the room and speakers are well matched, the center position should be right.

Before you balance the right and left channels, adjust and tweeter and midrange controls for tonal balance. Start by switching your receiver or amplifier into the mono mode and setting the bass and treble tone controls in the flat position. Rotate the balance control all the way to one channel so you can work on one speaker at a time. Turn the tweeter control, and the midrange pad if there is one, all the way down. Then advance the control until you hear a difference in the sound. When it is right, the highs should appear to blend with the bass. Don't overdo it.

After going through this procedure with each speaker, adjust the balance control until the sound source appears to be midway between the two speakers. For this test, have someone operate the control while you sit in your normal listening area on the center line of the speakers. When the channels are balanced, set your FM tuner or receiver to pick up interstation hiss and listen again. If the tweeter controls are properly balanced, the hiss should come from the same center position as the midrange sound. If one tweeter is set too high, the hiss will appear to be off center, toward the speaker with the stronger tweeter. Fine-tune the tweeter adjustment on that channel until the hiss is centered between the speakers.

Now listen to some music. If the sound is too dull, advance the tweeter controls and go through the whole procedure again.

After fine-tuning your speakers in this way, you might be surprised at their performance with the best of today's program material. Unfortunately, you will probably also become more aware of the mediocrity of most recordings. For a comparison of recording techniques, listen to "Brekkens Farm" or any other band on the Digital Music Products CD of "Neon" with Flim and the BBs. Then put on any random selection of a recording made by one of the large record companies and compare the naturalness of the sound.

## LISTENING: The Final Tweak

Various tweakers have reported that an amplifier, a speaker, and even cables need to be "run in" for a period of time to reach optimum performance. According to them, you should warm up your entire stereo system for a few hours before you do any critical listening.

My wife, Charys, has another theory on this. She says it is the listener who needs the "run-in," not the equipment. Whether or not she is right, your attitude *can* affect your enjoyment of the music. I once knew a group of audiophiles who got together about once a week to listen to each other's systems (they build their own speakers and amplifiers). I never witnessed them listening to a complete recording. The music would start with each one nodding in agreement to the sound, and then somewhere along the way someone would suddenly say, "Did you hear that?" Another would join in to say, "Yeah, stop the music." Then they began to tinker. The purpose of their informal group was to perfect a sound system, never to listen. In setting up a stereo system, there is a time to be critical and a time to enjoy.

# Chapter 4

# How to Build Enclosures

$A$ woofer out of its enclosure is as helpless as a fish out of water. And a poor enclosure can make a good speaker sound bad. Any careful amateur can build a box that is better in construction quality than the average commercial product.

Many people look on speaker enclosures as nothing more than furniture. Style and finish might be important to you and that's fine, but don't overlook the basic function of an enclosure. If the sound is bad, sooner or later it will begin to look bad.

### SPEAKER BOXES VS. MUSICAL INSTRUMENTS

You have probably heard that wood cabinets are best for their "mellow tone." The idea that wood makes good speaker enclosures is most likely a holdover from people's association with violins or pianos. However, it is a fallacy because a speaker system does not serve the same function as a musical instrument. An instrument is designed to produce a sound of a specific quality by adding certain overtones to the fundamental tone. Without overtones, it would produce uninteresting sine waves, and few listeners find an audio generator very musical. The speaker's role is different. Instead of producing original sound, the perfect speaker reproduces sound without altering it.

The speaker box is the silent partner of the drivers. The main acoustical function of an enclosure is to seal off the woofer's back wave from the front. Without it, the partial vacuum on one side of the moving cone would quickly cancel the pressure build-up on the other side and kill the sound. At low frequencies, this cancellation is almost total, so a bare speaker sounds "thin."

Tweeters and most midrange speakers have their own built-in enclosures. The sealed backs of these higher-frequency drivers protect them from the woofer's back wave. Open-back midrange speakers must have sub-enclosures to prevent distortion or even damage by the high internal box pressure.

Enclosure wall vibration is an evil; it robs bass and adds coloration. Several years ago, a booklet issued by a speaker company suggested that the enclosure didn't have to be strong. After all, the booklet stated, the cone of a speaker is just a piece of paper. Going on the analogy that a chain is no stronger than its weakest link, one might conclude that it's futile to build a sturdy enclosure. To refute that statement, all you have to do is to lightly place your finger tips on the surfaces of any speaker enclosure while the system is operated at a high level. You will undoubtedly feel some vibration. To see why, compare the size of the typical box to the size of a room. In most cases, the room is at least 1000 times the cubic volume of the box. The sound from one side of the cone fills the room; the sound from the other side is trapped in the box. It's no wonder you can feel vibration.

## ENCLOSURE SHAPE

A speaker box can add false sound in several ways. As mentioned above, the walls can vibrate. Another way is by the air resonances in the box.

Enclosures are usually shaped like rooms so they behave like rooms. In these little boxes, as in rooms, there are three kinds of resonances produced by the dimensions of the box: axial, tangential, and oblique. *Axial modes* are considered the most important, reflecting between opposite walls. For a six-sided box, there are three fundamental axial frequencies, one for each axis of the box. For a cube, the three axial modes are identical in frequency. (Bad.)

*Tangential modes* are formed by waves that bounce off four of the walls and move parallel to the other two walls. These modes have half the energy of axial modes.

*Oblique modes* are formed by waves that bounce around all six walls of the enclosure at various angles. They have only a quarter of the energy of axial modes.

It seems only prudent to try to prevent a really bad situation by using a dimension ratio that will provide some spacing of the standing wave modes in an enclosure. It is better to avoid coloration than to correct it.

TABLE 4-1 shows some preferred dimension ratios. These ratios refer to the internal box dimensions. The diagram shows an enclosure with greater depth than width. These width and depth ratios can be reversed if necessary, but deep enclosures usually perform better than shallow ones.

To plan an enclosure, first find the required cubic volume. This figure is usually supplied by the driver manufacturer, or you can use one of the design methods described in Chapters 5, 6, or 7. It is convenient to use cubic inches (in.$^3$), because most material is available in inches and even fractions of an inch. If the volume is stated in liters (l), remember that 1 cubic foot (ft.$^3$) equals 1728 cubic inches (in.$^3$), which equals 28 liters (l). After finding the desired cubic volume, select a tentative ratio from TABLE 4-1. Then go to TABLE 4-2 to find the intermediate dimension ($D_2$) for that cubic volume and dimension ratio.

Here is an example: Suppose you want to build an enclosure

**Table 4-I    Recommended Dimension Ratios for Speaker Enclosures**

| Ratio | $D_1$ (width) | $D_2$ (depth) | $D_3$ (height) |
|-------|-------|-------|-------|
| A | 0.79 | 1.00 | 1.26 |
| B | 0.80 | 1.00 | 1.25 |
| C | 0.62 | 1.00 | 1.62 |
| D | 0.67 | 1.00 | 1.80 |
| E | 0.62 | 1.00 | 1.44 |
| F | 0.80 | 1.00 | 1.20 |

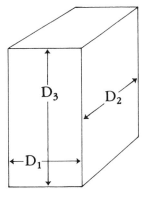

**Table 4-2  Intermediate Internal Dimensions for the Ratios in Table 4-1
for Various Cubic Volumes**

| $V_B$ (in.³) | A | B | C | D | E | F |
|---|---|---|---|---|---|---|
| | | | *Box Ratio* | | | |
| 3000 | 14.4 | 14.4 | 14.4 | 13.6 | 15.0 | 14.7 |
| 2500 | 13.6 | 13.6 | 13.6 | 12.8 | 14.2 | 13.8 |
| 2000 | 12.6 | 12.6 | 12.6 | 11.9 | 13.1 | 12.8 |
| 1800 | 12.2 | 12.2 | 12.2 | 11.5 | 12.7 | 12.4 |
| 1600 | 11.7 | 11.7 | 11.7 | 11.0 | 12.2 | 11.9 |
| 1400 | 11.2 | 11.2 | 11.2 | 10.5 | 11.6 | 11.4 |
| 1200 | 10.6 | 10.6 | 10.6 | 10.0 | 11.1 | 10.9 |
| 1000 | 10.0 | 10.0 | 10.0 | 9.4 | 10.4 | 10.2 |
| 800 | 9.3 | 9.3 | 9.3 | 8.8 | 9.6 | 9.4 |
| 600 | 8.4 | 8.4 | 8.4 | 8.0 | 8.8 | 8.7 |
| 500 | 7.9 | 7.9 | 7.9 | 7.5 | 8.3 | 8.2 |
| 400 | 7.4 | 7.4 | 7.4 | 7.0 | 7.8 | 7.6 |

with a cubic volume of 1600 cubic inches and you want to use ratio D for a tall narrow box. Table 4-2 indicates that the depth of the enclosure should be 11.0 inches. Returning to TABLE 4-1, the width ($D_1$) is 0.67 times 11.0 inches, or 7.40 inches. The height ($D_3$) is 1.80 times 11.0 inches or 19.80 inches. These internal dimensions — 7.4×11.0×19.8 inches, give an internal volume of 1612 cubic inches. To build a box, remember to add the thickness of two walls to each of those figures to get the outside dimensions.

You might have noticed that dimension ratios A through C are *constant ratios*, which means that the ratio of the largest dimension to the middle dimension is equal to the ratio of the middle dimension to the smallest. The theory of using constant ratios is that an enclosure so designed should provide equal geometric spacing of standing wave modes inside the box. But each of the other ratios have proponents. The last ratio (F), for example, is favored by some designers because it is the same ratio as 4:5:6 in whole numbers. No dimension is a multiple of the others. It produces a shorter, deeper enclosure than the others, using less material for a given cubic volume. If you have a special reason to use a ratio not shown here, go ahead. But try to avoid extreme shapes such as a cube.

## MATERIALS

The shape of the box is probably less important than its construction. Particle board has replaced plywood as the most popular choice for speaker enclosures. There seems to be some preference for medium-density fiberboard (MDF). High-density fiberboard, sometimes called industrial grade, is also used. There are conflicting views here. It takes more energy to move a high-density wall, but the greater the density and mass, the slower the energy is released again. This stretching of pulses can blur transients. There is some concensus to make small enclosures of high-density materials to put the resonant frequencies high enough for damping material to temper them. The methods used to stiffen small high-density walls also raise the frequency of resonance.

The traditional material for homemade enclosures—plywood—can be used if carefully selected. Cheap plywood has voids and loose layers that can rattle. Baltic birch plywood is good but expensive. Solid wood makes beautiful enclosures, but it must be lined with another material, such as fiberboard, because as it ages, solid wood can warp, shrink, or split. Any of these possibilities can compromise the integrity of the box by introducing air leaks.

For all but the smallest enclosures, ¾-inch material is desirable. If the box is large (more than a few cubic feet), the walls should be a full 1 inch thick. To obtain a 1-inch wall, glue a sheet of ¼-inch plywood to a piece of ¾-inch fiberboard (FIG. 4-1D) because combinations of different materials are usually better than a single material of the same thickness. One combination that works very well is a sandwich of a soft material between two layers of thin, hard material (FIG. 4-1E). The ultimate in this kind of construction is to use sand or other granular particles in the middle. The loose substance in the sandwich absorbs and dissipates energy, killing resonances.

You can improve almost any wall by deadening the inside of it with asphalt roofing material and/or foam-backed carpeting. But remember to allow for extra cubic volume when adding these materials because the interior dimension can be reduced by as much as ¼ to ⅜ inch per panel.

To apply one or more layers of asphalt roofing material, glue pieces of roofing to the central area of the panel (FIG. 4-1A). You can often get pieces of asphalt shingles free from the discard piles at building sites. Asphalt cement would be a logical choice for this job, but it takes days or even weeks to set. An all-purpose adhesive made by Macco, a division of Glidden, works well. It is sold under the trademark Liquid Nails. Whatever adhesive you use in a speaker enclosure, give it time to set before you install the speakers. The fumes from some adhesives can loosen the glue on driver cones.

After adding one or more layers of asphalt material, you can get

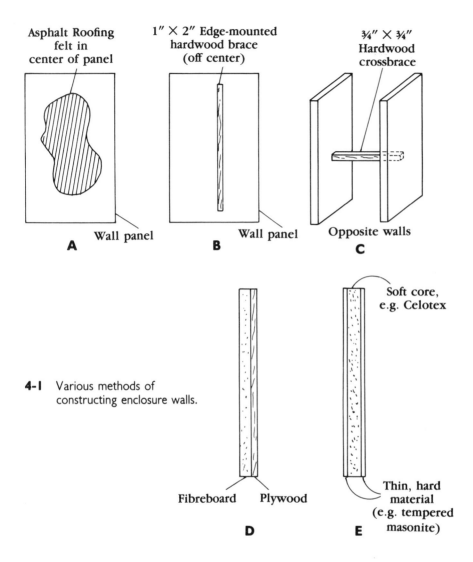

Asphalt Roofing felt in center of panel

1″ × 2″ Edge-mounted hardwood brace (off center)

¾″ × ¾″ Hardwood crossbrace

Wall panel

**A**

Wall panel

**B**

Opposite walls

**C**

Soft core, e.g. Celotex

Fibreboard    Plywood

**D**

Thin, hard material (e.g. tempered masonite)

**E**

**4-I**   Various methods of constructing enclosure walls.

additional damping by covering the interior of the box with a layer of foam-backed carpet. Use thin carpet with a backing of ⅛-inch polyurethane foam. Such carpet is cheap but does a good job without occupying much space.

Figure 4-1C is noteworthy. A ¾-×-¾-inch piece of hardwood or a 1-inch hardwood dowel can be used for crossbracing. Such a brace can be thin, because any movement in the walls tends to compress or stretch the brace the direction it is stiffest. If you use such a brace, it should be attached with glue and screws (best) or nails.

Two projects in this book, Projects 1 and 4, make use of high-density ceramic materials for improved wall rigidity.

## CONSTRUCTION RULES

Make your box assembly with screws and glue (FIG. 4-2). Ordinary carpenter's glue is good for close-fitting plywood parts, but epoxy is better for fiberboard. You can use ordinary wood screws, or if you have a power driver, wallboard screws work well. Wallboard screws are designed for rapid driving without pilot or guide holes. For better construction, take time to drill the holes.

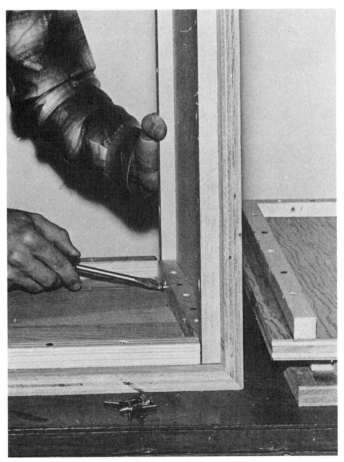

**4-2** If you don't have clamps, use screws to hold enclosure parts together while the glue is setting.

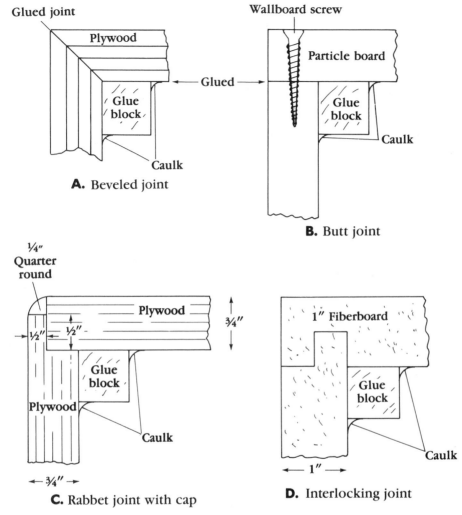

Glued joint

Plywood

Glue block

Caulk

**A.** Beveled joint

Wallboard screw

Particle board

Glue block

Caulk

**B.** Butt joint

¼" Quarter round

Plywood

¾"

½"  ½"

Glue block

Plywood

Caulk

← ¾" →

**C.** Rabbet joint with cap

1" Fiberboard

Glue block

Caulk

← 1" →

**D.** Interlocking joint

**4-3**   Various methods of making strong corner joints.

Figure 4-3 shows several ways to join panels at the corners of your enclosure. Glue blocks add to corner strength, particularly if they are installed with screws to make a tight fit while the glue is setting (FIG. 4-2).

If you use hardwood plywood for your enclosures, the joint shown in FIG. 4-3C deserves consideration because the strip of solid wood on the corner protects the edges of the plywood veneer. That joint is used for the upper corners of Project 5.

For particle board enclosures, the ordinary butt joint is a sensible

**4-4**  An H brace ties enclosure walls together, reducing vibration (Project 5).

choice with ¾-inch material. For 1-inch material, the interlocking joint shown in FIG. 4-3D is good.

Even with good material of adequate thickness, you will have some residual vibration. One way to increase the stiffness of the walls is to tie the walls together with an "H" brace as shown in the construction photograph of FIG. 4-4 (Project 5).

All joints should be caulked to prevent air leaks. Silicone rubber is a good choice for that and also for gasketing behind speakers.

## SPEAKER PLACEMENT AND MOUNTING METHODS

In building a compact enclosure, there is often very little choice on where to install the speakers. If you do have a choice, avoid putting the woofer at the midpoint of the cabinet height. The worst situation

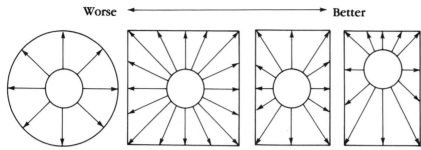

Worse ◄────────────────► Better

**4-5**  If the sound path length from the speaker to the edge of the baffle is equal in different directions, it will produce uneven frequency response.

would be a circular baffle with the speaker at the center (FIG. 4-5). There is some advantage in making the distance from each speaker center to the various edges of the box different. Place the drivers in a vertical line, if possible, and offset the line if there is space. Even a half inch off center will make for smoother high frequency response. For offset speakers, make stereo pairs in mirror images of each other.

Adjacent drivers should be mounted as close to each other as is practical. Theoretically, the distance between a woofer and tweeter in a two-way system should be no greater than one wavelength at the crossover frequency. To get the wavelength for any frequency, use the formula:

$$\lambda = \frac{13500}{f}$$

where $\lambda$ is the wavelength in inches
f  is the frequency in Hz

For example, if the crossover is put at 2000 Hz, the center-to-center distance for the drivers should be about 6¾ inches or less. This rule is often violated.

Nearly all modern speakers are designed to be installed from the front of the enclosure. If you can flush mount your speakers by one of the methods shown in FIG. 4-6, good. A router makes flush mounting easier, but careful work with a wood chisel made the installation shown in FIG. 4-7. If you don't flush mount, glue a layer of foam or felt to fill the space on the board to a depth that makes the foam flush with the driver frames. Carpet dealers sell a foam underlayment that works well here. You can buy it in various colors and thicknesses. Glue the smooth side of the foam to the board, leaving the soft side out.

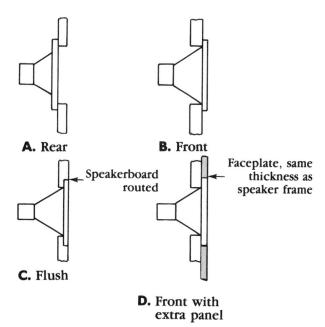

**A.** Rear      **B.** Front

Speakerboard routed →

→ Faceplate, same thickness as speaker frame

**C.** Flush

**D.** Front with extra panel

**4-6** Several methods of installing speakers. Methods C and D are preferred.

**4-7** This driver was installed by chipping out the plywood with a wood chisel for flush mounting.

You can use T-nuts for installation of woofers (FIG. 4-8). For temporary mounting, use a layer of foam weatherstripping behind the woofer.

## DAMPING MATERIAL

Don't forget to use damping material. Speakers that don't have enough damping material sound loud because of peaky response. Damping material in the box absorbs sound that would otherwise bounce around the box and back to the cone. Such reflections onto the cone either interfere or reinforce its output, causing peaks and dips in the speaker's response. Reflections from the back wall behind the woofer are the worst problem. It pays to use extra material in that area.

Damping material can help suppress modal behavior, but for greatest effectiveness, should theoretically be located at the points of highest pressure. For axial modes, that means away from the walls. There is an old rule of thumb that works well on this: small closed boxes should be loosely filled with stuffing.

Most builders put the damping material on the walls. In that location, it will have an cumulative effect on tangential and oblique modes. It might even be adequate for axial modes. In the days of huge speaker systems, it was common practice to cover the back, one side, and either the top or bottom interior surface. Theoretically that was enough, it was thought, to trap waves bouncing between opposite sides. You can often make a noticeable improvement in those enclosures by covering the other panels.

Perfectionists prefer to space the damping material a short distance from the walls. The ideal spacing depends on the density and thickness of the material. For fiberglass blankets of 1 to 2 inches in thickness, the typical spacing is ½ to 1 inch, with the wider spacing being used for the thinner material. This requires a set of spacers in the enclosure to hold the sheets of fiberglass or other material.

The traditional material for reflex enclosures is 1-inch acoustical fiberglass. Place it on the sides and back of the enclosure, but keep the opening to the port free unless it is a damped-port design.

Where the material is used to fill a box, don't choose fiberglass. There is a chance that fibers will get into the woofer assembly. Polyester batting is fine. Unlike fiberglass, it is relatively harmless to the speakers and to you.

For transmission line speakers, long fiber wool has been the favorite, but it has serious disadvantages. It is expensive and can harbor moths. Polyester seems more practical. One company, Ma-

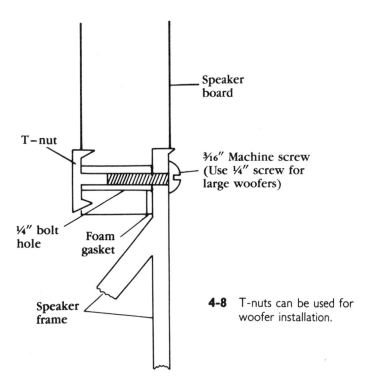

**4-8** T-nuts can be used for woofer installation.

hogany Sound of Mobile, Alabama, markets a special grade of synthetic material called Acousta-Stuf. It has crimped fibers that work well in transmission lines and aperiodic enclosures as well as for filling any small box.

## FINISHING YOUR SPEAKER

The finish that you put on the outside of your cabinets does more than make them look good. It also can add external damping. If you use fiberboard for the box, a plastic laminate veneer will improve the sound. Or, for wood, put on several coats of varnish or, even better, hard lacquer. The easiest way to make sure the wood stains and varnishes or lacquers you use are compatible is to stick to a single brand or at least buy them from a knowledgeable dealer.

Try not to use steel wool in finishing speaker enclosures. Steel strands can remain in or on the cabinet where the woofer's magnet can draw them into the voice coil. If you must use steel wool, carefully remove all traces of spent wool before installing the speakers.

## MORE BOX CONSTRUCTION HINTS

Every panel should fit against another panel or a cleat so that there are no gaps. Check the speaker board and back carefully for this. Fill all gaps with caulking material. Any unplanned air leak can degrade performance.

Even if you get every joint airtight, you can impair the performance of your system by careless introduction of speaker leads into the box. You can use mounting plates with push terminals or binding posts installed on them. The binding posts are the preferred method of terminating the speakers' wiring. For a two-terminal job, drill two holes, one for each terminal. Bring the speaker wires out through each hole and solder them to the internal lugs on the terminals. Then mount the plate, using silicone rubber as a gasket. Fill the holes in the back from inside the box with silicone rubber caulking material.

When your enclosure is finished, check the quality by rapping each panel with your knuckles. You should hear a dead, but relatively high-pitched sound. If a panel sounds like a drum, it isn't braced well enough. A rattle tells you something is loose. To precisely locate a rattle, strike the box here and there with a rubber hammer. The hammer makes a dull thump so you can hear the rattle and locate it.

Set the finished enclosure on the floor first on one side and then another. In each position shake some powdered chalk on the panel facing up. Play some loud music and watch the chalk. The chalk, and your finger tips, can tell you where more bracing is needed. Even better, use an audio generator for this, as described in Chapter 10.

### PROJECT 1: Ceramic Tile Speaker Enclosures

This is two projects in one because you have a choice of making a 10-liter (⅓ cubic foot) box or a larger, 14-liter (½ cubic foot) model. Almost any 6½-inch woofer that is designed for closed-box duty will work well in one of these enclosures. For the SEAS P17RC, the manufacturer recommends a range of closed-box volumes from 7 to 18 liters. The larger enclosures produce a more highly damped system that some listeners prefer. A second choice of woofer is listed here (TABLE 4-3), and there are other possibilities of kits available from various sources that can be used in one of these enclosures. For example, the November and December 1988 issues of *Audio* carried a two-part article by Ken Kantor called "Speakers By Design." The article described a computer-designed system that should work well with the 10-liter box here.

Be aware that a slightly different method of construction is used

**Table 4-3  Components Lists for Two Versions of Project I**

*SYSTEM A*

| | | |
|---|---|---|
| 1 | 6½-inch woofer | SEAS P17RC |
| 1 | Tweeter | Polydax TW74A |
| 1 | 4 uF capacitor, Mylar, 100 V | C1 |
| 1 | 4 $\Omega$ resistor, 15 W | R1 |
| 1 | 10 $\Omega$ resistor, 10 W | R2 |
| 1 | Speaker terminal plate | |

SYSTEM B

| | | |
|---|---|---|
| 1 | 6½-inch woofer | Peerless 1744 |
| 1 | Tweeter | Polydax TW74A |
| 1 | 2.7 uF capacitor, Mylar, 100 V | C1 |
| 1 | 12 uF capacitor, NP electrolytic, 100 V | C2 |
| 1 | 0.7 mH inductor, air core | L1 |
| 1 | 0.3 mH inductor, air core | L2 |
| 1 | 4.5 $\Omega$ resistor, 15 W | R1 |
| 1 | 10 $\Omega$ resistor, 10 W | R2 |
| 1 | 8 $\Omega$ resistor, 10 W | R3 |
| 1 | Speaker terminal plate | |

for the two sizes of enclosures. If you look carefully at the photograph of the 10-liter boxes in Chapter 2, FIG. 2-1, and at that of the 14-liter boxes in FIG. 4-9, note that the larger boxes have rounded corners. They were made with a strip of quarter-round vinyl at each upper corner, as shown in detail B in FIG. 4-10. If you decide to build the 14-liter box but don't want to purchase the special quarter-round vinyl, you can use the same dimensions for all parts except the top, which must be wider by ½ inch. And, you will have to add some material to the top of the inner box to raise the top tile a bit; the thickness of that added piece should be ⅛ inch. To use the quarter-round vinyl on the smaller box, reduce the width of the top tile by ½ inch and shorten the inner box by ⅛ inch.

Before cutting the tile, lay out the pieces the way they will fit on the enclosure to make sure the grain, if any, is consistent. Even subtle patterns should be respected or the visual effect can be jarring. Any grain pattern should run vertically on the sides and continue across the top.

## Construction Procedure

Cut out the insulation board parts to the dimensions listed in TABLE 4-4. TABLE 4-5 shows the dimensions of the finished enclosures, as

**4-9**  A pair of 14-liter Ceramic Tile Speaker Enclosures (Project 1). Note use of quarter-round vinyl strips in upper corners.

### Table 4-4   Enclosure Parts Lists for Two Versions of Project I

| Number of Pieces | 12" Tile $V_B = 10 \ 1$ | $13^3/_{16}$" Tile $V_B = 14 \ 1$ | Function |
|---|---|---|---|
| *½" Insulation board* | | | |
| 2 | $8\frac{7}{8} \times 10^{11}/_{16}$" | $9\frac{3}{4} \times 11^{11}/_{16}$" | Sides |
| 2 | $8\frac{7}{8} \times 8$" | $9\frac{3}{4} \times 8\frac{3}{4}$" | Top & bottom |
| 2 | $8 \times 11^{11}/_{16}$" | $8\frac{3}{4} \times 12^{11}/_{16}$" | Front & back |
| *⅜" Plywood* | | | |
| 1 | $8 \times 11^{11}/_{16}$" | $8\frac{3}{4} \times 12^{11}/_{16}$" | Speaker board |
| 1 | $8\frac{3}{4} \times 12\frac{1}{2}$" | $9\frac{1}{2} \times 13\frac{1}{2}$ | Grille frame |
| *Ceramic tile* | | | |
| 2 | $10\frac{5}{8} \times 12$" | $11\frac{9}{16} \times 13^3/_{16}$" | Sides |
| 1 | $10\frac{5}{8} \times 7\frac{3}{4}$" | $11\frac{9}{16} \times 8\frac{1}{2}$" | Bottom |
| 1 | $10\frac{5}{8} \times 8\frac{3}{4}$" | $11\frac{9}{16} \times 8\frac{1}{2}$" | Top |
| 1 | $7\frac{3}{4} \times 11\frac{1}{2}$" | $8\frac{1}{2} \times 12^7/_{16}$" | Back |
| | Dimensions for 10 l enclosure are for corner A in FIG. 4-10. | Dimensions for 14 l enclosure are for quarter-round vinyl in upper corners. *See text.* | |

**Table 4-5   Enclosure Dimensions for 10-Liter and 14-Liter Versions of Project 1**

| Dimension | 10 liter | 14 liter |
|---|---|---|
| EW | 8¾″ | 9½″ |
| EH | 12½″ | 13½″ |
| ED (excluding grille) | 10⅝″ | 11⁹⁄₁₆″ |
| IW | 7″ | 7¾″ |
| IH | 10¹¹⁄₁₆″ | 11¹¹⁄₁₆″ |
| ID | 8⅞″ | 9¾″ |

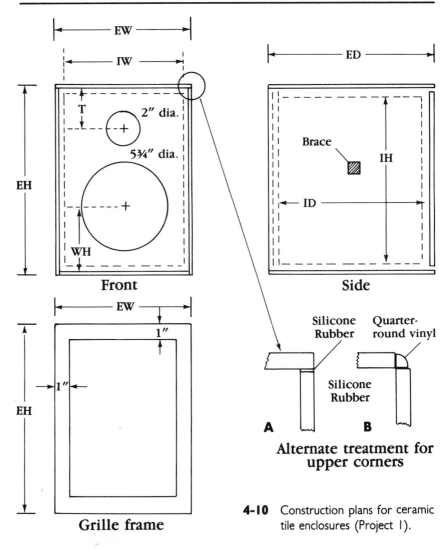

**4-10**  Construction plans for ceramic tile enclosures (Project 1).

shown in FIG. 4-10; however, the internal dimensions shown there do not take into account the asphalt roofing liner. To assemble the interior box, use four-penny finished nails and Liquid Nails adhesive. The nails go down through the top and up from the bottom into the sides. Glue and clamp the plywood speaker board to its soft-board base. Spread enough Liquid Nails on the insulation board to make a squishy feel when you twist the plywood about, but not enough to hold the boards apart. Clamp or screw the softboard to the plywood and leave it until the adhesive sets. Next, mark and cut out the speaker holes. If you use different drivers from the ones listed here, you might have to adjust the size or position of the holes to match your speakers. After cutting out the holes, set the speakers in the holes to see if they fit. Paint the speaker board black.

Glue and nail the combination speaker board to the shell of the enclosure (FIG. 4-11). Cut a ¾-×-¾-inch hardwood brace to fit between the side walls of the box. Place the brace near the center of the walls, just off center is ideal, and glue and screw the walls to the brace. Glue and staple pieces of asphalt roofing to the interior of the sides, top, and bottom panels. Remember that you should use Liquid Nails for all gluing of asphalt materials such as insulation board and pieces of roofing unless you use the much slower-setting asphalt cement.

Now for the most unusual procedure in this project — cutting the ceramic tile. Tile layers use a special tool for this, and you might be able to hire someone to do the job. I cut my tile with a masonry blade in my power saw. If you do that, wear goggles and a dust mask and make sure you have good ventilation in the area where you work.

To cut ceramic tile with a masonry saw blade, set the blade for a shallow cut and gradually increase the depth of the cut until the blade is about halfway through the tile. Then flip the tile and make cuts from the other side until the piece is cut through. One detail of cutting the top piece that should be considered is that most tile comes with slightly rounded edges. For identical side edges on the top piece, you must make two cuts from front to back, one for each edge.

When I planned my enclosures, I put the terminal strip in the center of the back (FIG. 4-9). If you put it near the bottom, it is easier to apply asphalt liner to the central area of the back. I also suggest a different approach to terminal installation. Instead of drilling ⅜-inch holes with a carbide-tipped masonry bit, you can use the masonry saw to remove notches at the bottom of the panel for wire leads. With

**4-11**   Inner boxes of Project 1.

some kinds of tile, hole drilling can be tedious. If you decide to drill the holes rather than to cut notches, drill from the exterior, glazed, surface. When I bought a carbide bit, the salesman told me to drill tile from the unglazed side, but when I tried to do that, the bit chipped out a large area in the glazed surface as it went through.

After notching or drilling the back, glue the insulation board back piece to the inside surface of the tile back. When you clamp the pieces together, make sure the tile is centered on the slightly larger piece of softboard. When the glue is set, extend the notches, or holes, through the insulation board.

Mount the terminal hardware and internal speaker wiring to the back. As you may have noticed, FIG. 2-1 shows a four-terminal strip; FIG. 4-9 shows a three-terminal strip. By separating woofer and tweeter inputs, you can experiment with crossover wiring from outside the box. Or, the double terminals can be used for bi-wiring.

The crossover network diagram for the SEAS woofer is shown in FIG. 4-12A, and that for the Peerless woofer is in FIG. 4-12B. If desired, you can substitute an 8-ohm L-pad for resistors R1 and R2 in either circuit. Inserting the L-pad will provide more versatile control of

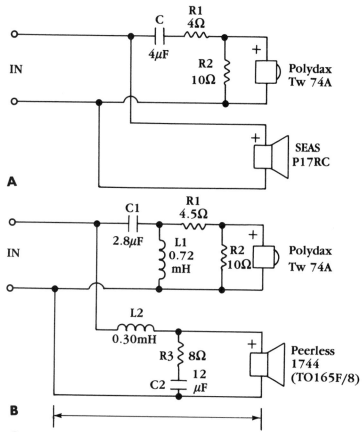

**4-12** Crossover network diagrams for two versions of Project I.

woofer/tweeter balance, but you must drill an extra hole in the tile back plus a larger hole in the insulation board back to countersink the L-pad onto the tile. For this you will need an L-pad that will mount in a ⅜-inch hole.

If you are using one of the woofers listed here, you can wire the crossover network and install it at this time. Mount it on the inside of the back panel, near the speaker terminals.

Attach the composite back to the box with adhesive. Note that Liquid Nails is the adhesive for use with any gluing that involves insulation board. Apply weight to the back while the glue sets. Next, invert the box and spread adhesive on the bottom. Push the bottom tile piece onto the adhesive, twisting it slightly to spread the glue over the surface. Don't move the box until the glue has thoroughly set. Follow this same procedure for each of the other tile parts (FIG. 4-13).

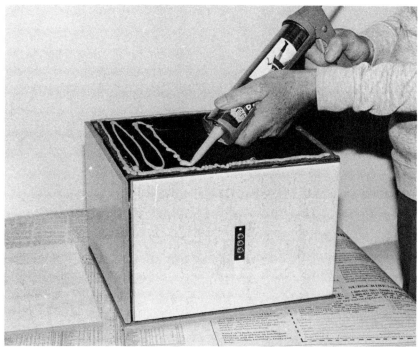

**4-13** Apply Liquid Nails adhesive to insulation board panels, then press down ceramic pieces and twist slightly to spread glue evenly under the tile.

When all the ceramic panels are installed, fill the spaces between them with silicone rubber cement. I used white cement to match the off-white tile on my enclosures. As designed, no two pieces of tile touch each other. They are separated by a strip of flexible silicone rubber.

If you didn't install the crossover network earlier, it can be inserted through the woofer hole. I mounted mine on a small board that I glued inside the back panel after assembly of the box. Then I filled the box with polyester stuffing. Don't force too much damping material into the enclosure; use just enough to loosely fill it.

When you wire the speakers, don't forget to observe polarity. Run a thin bead of silicone rubber sealant under the tweeter to gasket it. For temporary testing, you can seal the woofer with a gasket of 3/16-×-3/8-inch foam weatherstrip tape. Use 1/2-inch screws to hold the woofer in place for testing. For final installation, set the box on its back and run a bead of silicone rubber around the cutout. Place the woofer on the bead of sealant and twist it slightly to make sure the gasket is complete. Leave the speaker in that position for a day before using your speakers.

To reduce diffraction, glue a piece of ⅜-inch foam carpet under-layment on the baffle. Cut a piece of foam to fit within the grille frame; then make cutouts for the speakers and glue onto the speaker board. To hold the grille frame in place, countersink four pieces of matching velcro on the speaker board and the back of the grille frame. As a covering for the grille frame, any thin polyester knit material that you can see through when you hold it up to the light should work well.

## PROJECT 2: Mini Speakers

Most mini speakers are good for a dorm room or other secondary duty, so don't expect them to have full bass. The most you can hope for is clean sound at moderate volume. The speakers in this project offer that in a pair of enclosures that can fit in just about any small space.

The Audax HD 10 P FSC woofer used here has such extended range that the tweeter is almost unnecessary. The main reason for adding a tweeter is to get better high-frequency dispersion.

Specifications for this woofer list a resonance frequency of 52 Hz and a total Q of 0.4. The speakers I got had much higher values for each of these characteristics. You can substitute just about any available 4-inch woofer.

### Table 4-6   Parts List for Project 2: Mini Speakers

| Number of Pieces | Dimensions | Function |
|---|---|---|
| 2 | ½ × 7 × 8½″ plywood | Sides |
| 2 | ½ × 7 × 5¾″ plywood | Top & bottom |
| 2 | ½ × 4¾ × 7½″ plywood | Front & back |
| 1 | ¼ × 8½ × 5¾″ plywood | Grille frame |
| 6 ft. | ½ × ½″ hardwood | Cleats & glue blocks |
| 1 | ¼ × 3½ × 3½″ masonite | L-pad board |
| | Components | |
| 1 | 4-inch woofer | Polydax HD10PFSC or other |
| 1 | Tweeter | Polydax TW 74 A |
| 1 | 3.3 uF capacitor, Mylar, 50 V | C |
| 1 | 8 Ω L-pad, 15 W | Tweeter control |
| 1 | Speaker terminal strip | |

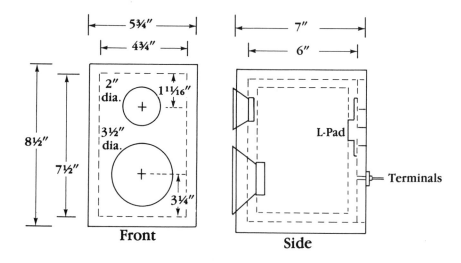

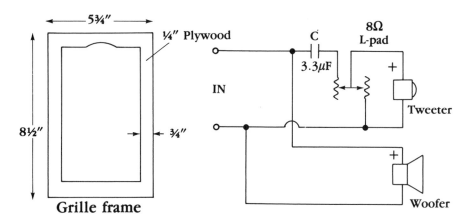

**4-14** Enclosure plans and wiring diagram for Project 2.

For the box, I used nine-ply ½-inch Baltic birch plywood of very high quality. In such a small enclosure, it is quite satisfactory. Follow the instruction procedures described in this chapter. The enclosure details and crossover diagram are shown in FIG. 4-14. Use ½-×-½-inch cleats behind the front panels, glue blocks in the corners, and caulking at every joint just as in a big box. Mount the 15-watt L-pad on a small piece of tempered masonite and install it on the upper part of the back panel (FIG. 4-15 and FIG. 4-16). I glued and stapled thin ¼-inch carpet, as recommended in this chapter, on the internal walls with the ⅛-inch polyurethane foam backing against the wood wall. I covered the back with a 1-inch piece of fiberglass, then loosely filled the box with polyester stuffing.

If you don't expect too much in the way of bass response, there's almost no way you can go wrong with this little project.

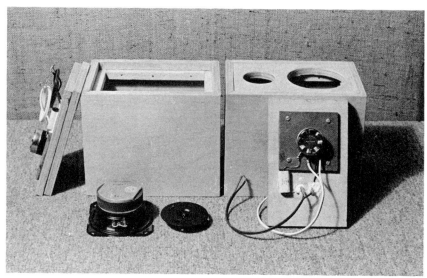

**4-15**   Drivers and enclosure parts for Project 2.

**4-16**   Front and rear view of Project 2: Mini Speakers.

*Chapter* **5**

# Closed-Box Speaker Systems

$S$hould you like to do things the easy way, closed-box speakers could be your thing. They are simple to design, easy to build, and they don't change significantly in performance as the woofer ages.

Design charts show that closed-box speakers are inferior in bass range to reflex speakers of the same cubic volume. In real terms, this advantage for reflex speakers is debatable. The closed-box speaker cuts off at a much slower rate than the reflex, losing 12 dB per octave below resonance instead of 24 dB per octave. This means that a closed-box speaker with the same theoretical cutoff frequency as a reflex might give the impression of lower bass. But the point about the 12 dB per octave cutoff that some closed-box fans make is that a sharp cutoff system always has more tendency to ring than a system with a gentle slope. Designing a closed-box speaker is primarily an exercise in choosing the right cubic volume for your driver.

## BOX SIZE AND WOOFER PERFORMANCE

When you install a woofer in a box, the pressure inside the box acts on the cone as a restoring force, or stiffener. The smaller the box, the stiffer the air in the box to the woofer. This stiffening process raises the frequency of resonance and the Q of the speaker (FIG. 5-1).

For a box of given cubic volume, the larger the driver cone, the greater the change of resonance frequency and Q. The illustration of

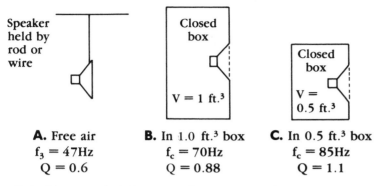

**A.** Free air
$f_3 = 47$Hz
$Q = 0.6$

**B.** In 1.0 ft.$^3$ box
$f_c = 70$Hz
$Q = 0.88$

**C.** In 0.5 ft.$^3$ box
$f_c = 85$Hz
$Q = 1.1$

**5-1**   How a speaker's frequency of resonance and Q vary in free air and in boxes of different cubic volumes.

FIG. 5-2 shows how two woofers with the same free-air resonance behave when placed in boxes of equal size. The curve for area $A_1$ represents a woofer with a larger cone area than the woofer represented by $A_2$. Note that as the box volumes are made smaller and smaller, the closed-box resonance frequencies climb sharply, showing the effect of pressure build-up behind the cone. In fact, the closed-box resonance ($f_c$) for a closed-box speaker system varies inversely with the square of the cone area. That is why large speakers demand large boxes.

To see how the low-frequency response of an individual speaker varies when put into boxes of different sizes, look at the graphs in FIG. 5-3. The numbers on the curves show the $Q_{TC}$, or closed-box Q, for the speaker in each box. The smaller the box, the higher the Q. If the driver is installed in a box that is too small, considerable low-bass range will be lost even though the speaker has a boomy midbass response. At the other extreme, a box large enough to produce a system with a Q of 0.5 has extended low bass — but at a low SPL compared to the midrange. The curve for a Q of 0.7 shows the flattest response.

Which curve is best? The answer to that depends on who is designing the speaker. Many commercial speakers have been designed for a Q of about 1 on the grounds that it provides bass at a level that is even with the midrange with only moderate peaking above cutoff. This curve is recommended by some engineers for small woofers to balance their limited low-bass range. Other designers using the same kind of woofer would choose the flatter curve of 0.7 for purity of tone in the middle bass. High-end speakers are often designed for a Q of 0.6 to 0.7 for higher damping.

As mentioned earlier, closed-box speakers are not as finicky

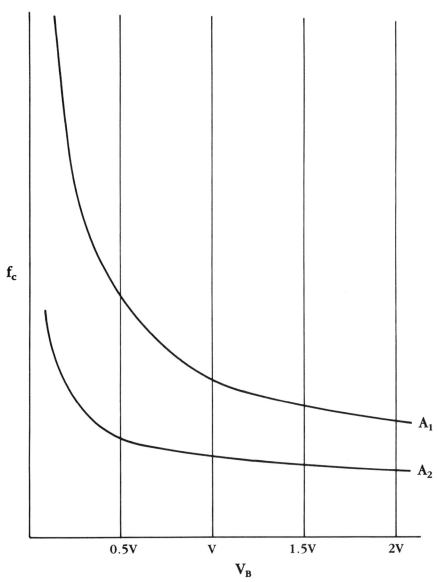

$f_c$

0.5V  V  1.5V  2V

$A_1$

$A_2$

$V_B$

**5-2**  How box volume affects resonance for two speakers. The cone area of $A_1$ is greater than that of $A_2$.

about design criteria as ported systems. If you have no specifications on a woofer and no way to measure it, you can use the chart in FIG. 5-4 to make an estimate of box volume. The shaded area in the chart shows the normal range of cubic volumes that should work with your speaker.

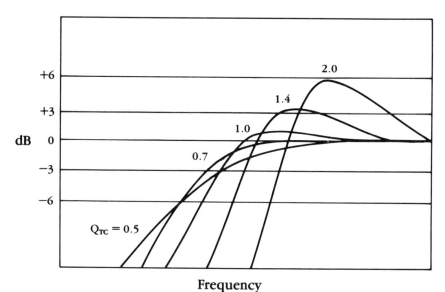

**5-3**  How a speaker's response changes when operated at various values of $Q_{TC}$. Each line represents a different $Q_{TC}$.

Some manufacturers publish a range of box volumes for each of their drivers. Others do not. If you have a woofer and know its specifications but have no box recommendation, you can easily calculate a suitable cubic volume.

## HOW TO APPLY TEST DATA TO BOX SIZE

To calculate the right box volume you need to know three specifications for your woofer. These are:

$f_s$ – the speaker's free-air resonance
$Q_{TS}$ – the speaker's total Q factor
$V_{AS}$ – the speaker's compliance stated as equivalent air volume

When you have these three figures, check the ratio of $f_s/Q_{TS}$ (F ratio) for your speaker. It is a significant ratio that can tell you whether your driver is best suited to a closed box or another type of enclosure. The value of the ratio is the resonance frequency of the driver in a closed box with a $Q_{TC}$ of 1. For large woofers, the F ratio should be in the neighborhood of 50 for best performance in a closed box. If it is 100 or higher and you like a pronounced bass response, consider a reflex enclosure. For a small woofer, the ratio is likely to be in the high range because of its higher $f_s$.

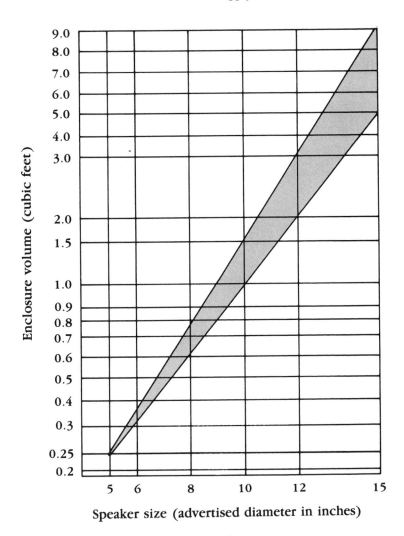

**5-4** Typical closed-box volumes for speakers of various diameters.

As an example of how two models of one brand of woofers vary in the F ratio, consider the Peerless 1510 and the Peerless 1556. The 1510 woofer has a resonance frequency of 37 Hz and a $Q_{TS}$ of 0.61, for an F ratio of 61. The 1556 has a resonance frequency of 29 Hz and a $Q_{TS}$ of 0.32, for an F ratio of 90. The two woofers have the same dimensions; the chief difference is that the latter has almost twice the magnet weight of the first. As you might expect from the F ratios, the

1510 is designed primarily for use in a closed box and the 1556 for a reflex.

If you want to design a closed box for your woofer, begin by choosing the value of $Q_{TC}$, or system Q, that you want. Then find the value of alpha ($\alpha$) that will give you that Q. Alpha is the ratio of $V_{AS}$ to $V_B$, where $V_B$ is the volume of the box you will build. You can find alpha by:

$$\alpha = (Q_{TC}/Q_{TS})^2 - 1$$

After finding alpha, you can calculate $V_B$ by:

$$V_B = V_{AS}/\alpha$$

Here is an example. Suppose you have a 6½-inch woofer with an $f_s$ of 37 Hz, a $Q_{TS}$ of 0.33, and a $V_{AS}$ of 36 liters. This has an F ratio of 112, which is rather high for closed-box design but not unusual for a small woofer. Also, a check with literature from the manufacturer (SEAS) shows that it is intended mainly for closed-box designs with recommended enclosure volumes of from 7 to 18 liters.

If you decide to aim for a $Q_{TC}$ of 0.7, what cubic volume should you choose? Using the formula:

$$\alpha = (0.7/0.33)^2 - 1$$

$$= 4.5 - 1$$

$$= 3.5$$

So the required box volume is:

$$V_B = 36\ 1/3.5$$

$$= 10.3\ 1 \text{ or } 0.37\ \text{ft.}^3$$

The example above was the basis for design of the 10-liter enclosure of Project 1 in Chapter 4.

If you want to estimate the system resonance ($f_c$), you can find it by using the same ratio as the Q ratio. For the example above:

$$f_c = Q_{TC}/Q_{TS} \times f_s$$

$$= 2.12 \times 37 \text{ Hz}$$

$$= 78 \text{ Hz}$$

**Table 5-1 Cutoff Frequency (f₃) for Closed-Box Speakers in Relation to System Resonance (f꜀)**

| $Q_{TC}$ | $f_3$ Factor |
|---|---|
| 0.5 | 1.55 |
| 0.6 | 1.21 |
| 0.7 | 1.0 |
| 0.8 | 0.9 |
| 0.9 | 0.83 |
| 1.0 | 0.79 |
| 1.1 | 0.76 |
| 1.2 | 0.74 |
| 1.3 | 0.72 |
| 1.4 | 0.71 |

For the theoretical cutoff frequency $(f_3)$—the frequency where the response is down 3 dB—you can find the multiplier in TABLE 5-1. Then:

$$f_3 = f_c \times \text{multiplier}$$

For an $f_c$ of 78 Hz and a Q of 0.7:

$$f_3 = 78 \text{ Hz} \times 1$$
$$= 78 \text{ Hz}$$

This is a theoretical figure. The real cutoff is probably lower in frequency, depending on room position. If, after making the calculations you feel that $f_3$ is too high, substitute another value for $Q_{TC}$ in the equations.

## PLOTTING YOUR WOOFER'S RESPONSE

Note the chart in FIG. 5-5. The curves in FIG. 5-3 were placed the way they would appear on a logarithmic scale; those in FIG. 5-5 are displaced further from each other for easier reading. With a few calculations, you can assign frequencies to the vertical lines at A, B, and C that fit your woofer's characteristics. As you can see, the curves for three values of $Q_{TC}$ cross the $-3$ dB line at the three vertical frequency lines. To fit the chart to your woofer, use these formulas:

For line A:   $f = f_s \times 0.7/Q_{TS} \times 1$

For line B:   $f = f_s \times 1.4/Q_{TS} \times 0.71$

For line C:   $f = f_s \times 2.0/Q_{TS} \times 0.67$

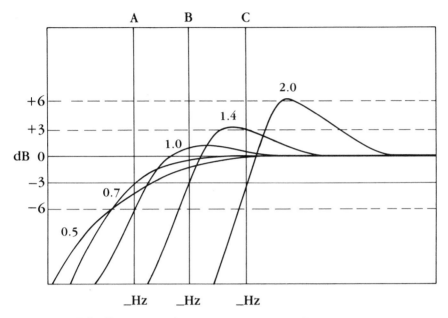

**5-5**   Design chart for closed-box speakers. (See text.)

For example, suppose you have a woofer with an $f_s$ of 24 Hz and a $Q_{TS}$ of 0.45. Working through the equations above, the three frequencies are: A = 37 Hz, B = 53 Hz, and C = 71 Hz. Knowing what frequencies the three lines represent for your woofer allows you to get a picture of what kind of bass performance is theoretically possible with different values of $Q_{TC}$. Furthermore, you can calculate the required box volumes for each $Q_{TC}$ and balance your desire for a certain kind of bass performance against the space needed by the enclosure for that performance.

## APERIODIC ENCLOSURES

Some designers view the closed box as a nonlinear pressure box. They suggest that there is an advantage in a pressure relief vent. This is the basis of the aperiodic design: a box with a small damped port. It is filled with a resistive material such as fiberglass or polyester stuffing.

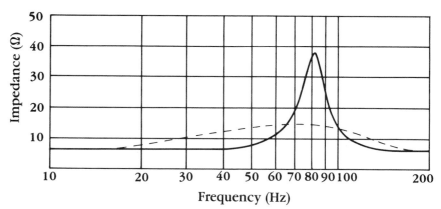

**5-6** Impedance curves for a speaker in a closed-box (solid line) and aperiodic enclosure (dashed line).

Figure 5-6 shows the impedance curve of an 8-inch woofer in a closed box (solid line) and after a damped port was added to the box (dotted line). Note that these are not frequency response curves. Some speaker authorities make much of a flattened impedance curve; others say it is irrelevant. Advocates of the aperiodic system compare it to the much larger transmission lines. They say the damped aperiodic enclosure avoids the possibility of ringing with a reflex while relieving the "oil can" pressure effect of the closed box.

To build an aperiodic enclosure, simply cut a vent and stuff it with damping material. The area of the vent should be about 10 square inches per cubic foot of enclosure volume. Here is one way to add a resistive vent. For an enclosure with ¾-inch walls, cut the vent and staple a piece of ¼-inch hardware cloth (coarse screenwire) across the interior of the vent. When the enclosure is completed, you can stuff the vent until the stuffing material is compressed and staple a piece of hardware cloth over the hole. The vent can be located in the back if desired. If you have test equipment, adjust the amount of damping material for the flattest impedance curve.

## MULTIPLE WOOFER SYSTEMS

If two woofers are installed in the same enclosure in the conventional manner, as in Project 11 in Chapter 9, the enclosure volume must be double that used for a single woofer. Figure 5-7 shows several unusual ways to build compound woofer systems. Each of these permits the outside woofer to perform as if it were on an infinite baffle. The two woofers can be wired in parallel or in series,

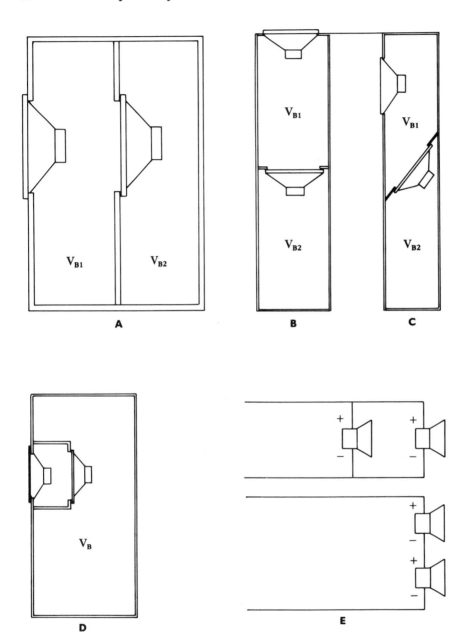

**5-7**   Several ways to install double-woofer closed-box speakers. In each of these, the exterior woofer behaves as if it were on an infinite baffle.

but in all the combinations shown in FIG. 5-7, they must be wired with the same polarity. And the woofers must be identical.

Figure 5-7D is an extreme example of the system. The tunnel that connects the two woofers can be as short as possible, just long enough to keep the magnet of the front woofer from hitting the rear cone. This system is sometimes called an "Isobarik" system. It was patented several years ago by Ivan Tiefenbrun, who produces various audiophile components in Scotland under the name of Linn Products. The name Isobarik comes from the term that means equal pressure. The pressure is made equal because the buried woofer moves in the same direction as the outside woofer, maintaining almost constant volume in the space between them.

Enthusiasts of this system say that the outside woofer is able to respond very much like a transmission line woofer down to its resonance frequency. What is more, the volume of the box in FIG. 5-7D can be made smaller than one would normally want to use for a single woofer because the two act as if cone stiffness is doubled, reducing $V_{AS}$. Unlike FIG. 5-7D, the systems in the plans shown in FIGS. 5-7A, 5-7B, and 5-7C have two equal volumes, each volume the same as that for a single woofer. Where space is at a premium, D is a logical choice.

Figure 5-8 shows still other ways to keep the woofer factories at full production. In these designs, you must wire the woofers with reverse polarity because they operate in a push-pull combination. The push-pull action cancels distortion. The extra woofer also has the advantage that it can be wired so that it will roll off its output in the upper bass or midrange to compensate for the lightweight sound of small speakers. There is more about this in Chapter 9.

## PROJECT 3: General-Purpose Speakers

This project follows a standard formula for a low-cost speaker, an 8-inch woofer, and a 1-inch dome tweeter in a simple box (FIG. 5-9). The enclosure plan is shown in FIG. 5-10, and the crossover diagram is in FIG. 5-11. In the interest of low cost and ease of wiring, no inductance is used in the woofer circuit.

The list of parts for Project 3 is given in TABLE 5-2. Follow the standard construction procedure as outlined in Chapter 4 and shown in FIG. 5-12. The only unusual feature is the 3-×-3-inch damped port that turns this into an aperiodic design.

I tried four layers of ⅜-inch polyurethane foam in the port compressed to a final thickness of ¾ inch by the outside hardware cloth. With that, the sound was somewhat cramped. Next, I placed 2

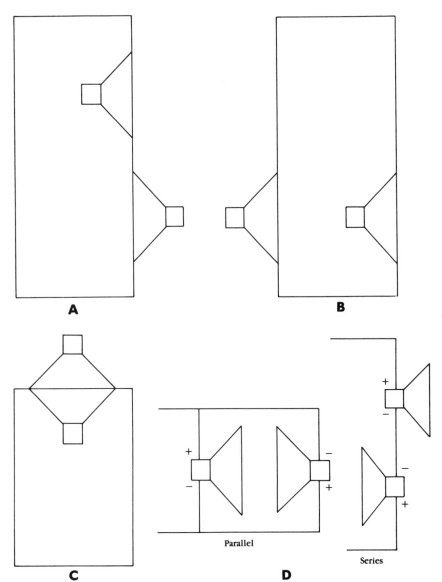

**5-8**   (A), (B) and (C) show three ways to install a push-pull closed-box system. (D) Shows the wiring scheme-either parallel or series. Note the woofers must be wired out of phase for push-pull operation.

inches of fiberglass in the port and compressed it to ¾ inch. The performance improved, but there was a suggestion of boxiness. With only 1 inch of fiberglass in the vent, the sound was more free. During all these tests I also varied the amount of damping material in the

**5-9**   Project 3: General-Purpose Speakers, with the grille cloth removed.

box, trying to find an optimum combination of material in the box and the vent.

My next experiment was to remove the fiberglass in the vent and replace it with about ¼ ounce of Acousta-Stuf. This appeared to improve the sound. My final arrangement was about 5 ounces of Acousta-Stuf in the box and just enough in the vent to make for a slight compression when the exterior hardware cloth was installed. Hence, this is a project that you can tailor to your own listening tastes by varying the amount of damping material.

## PROJECT 4: High-Quality Small Speakers

Flue tile might appear to be a strange material for a speaker enclosure, but it's not new. Several years ago, I tried it and liked the performance. When I moved, I gave the speakers to some friends. When they told me recently that their audiophile son had upgraded the old enclosures with new drivers and that they outperformed

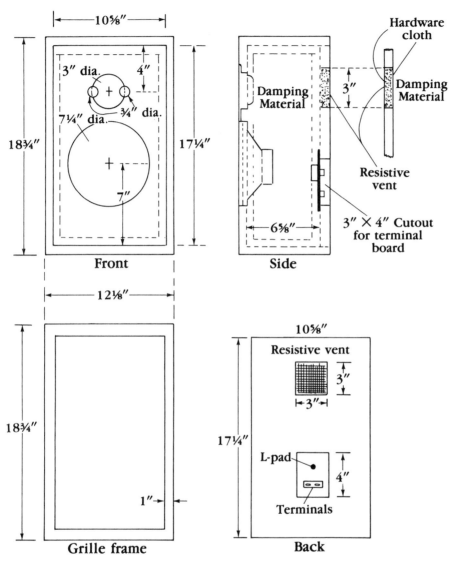

**5-10**  Construction plans for Project 3 enclosures.

every speaker in his collection, I decided to try it again with the speakers shown in FIG. 5-13.

These speakers are designed to have a $Q_{TC}$ below 0.7 for a lean but pure bass response. Low $Q_{TC}$s are often associated with speakers designed mainly for classical music. The speakers described here are at their best with classical and jazz artists. They can even be used for rock music, but the sound level is limited by the size of the woofers.

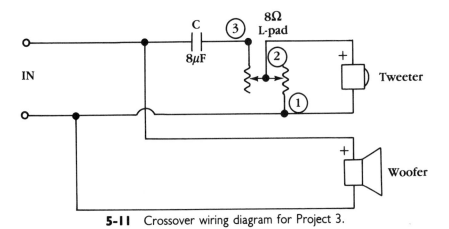

**5-11**  Crossover wiring diagram for Project 3.

**Table 5-2  Parts List for Project 3, General-Purpose Speakers**

| Pieces | Dimensions | Function |
|---|---|---|
| 2 | ¾ × 8⅛ × 18¾" plywood | Sides |
| 2 | ¾ × 8⅛ × 12⅛" plywood | Top & bottom |
| 2 | ¾ × 10⅝ × 17¼" particle board | Front & back |
| 1 | ⅜ × 12⅛ × 18¾" plywood | Grille frame |
| 8 ft. | ¾ × ¾" hardwood | Cleats & glue blocks |
| 1 | ¼ × 5 × 6" masonite | Control board |
| 2 | 5 × 5" hardware cloth, ¼" | Vent covering |
| 3 oz. | Polyester fill or Acousta-Stuf | Damping material |
| | **Components** | |
| 1 | 8" woofer | Pioneer polypropylene PB20 |
| 1 | 1" soft dome tweeter | Polydax HD12X9D25 |
| 1 | 8 Ω L-pad, 50 W | Tweeter control |
| 1 | 8 μF capacitor, 100 V | C |
| 1 | Speaker terminal plate | |

The parts list appears in TABLE 5-3. Flue tile comes in various sizes that are labeled 8 × 8, 9 × 13, 13 × 18 inches, and so on. It isn't made to precise dimensions. I bought a piece of 9-×-13-inch tile that measured about 8 × 13 × 23½ inches. Figure 5-14 and TABLE 5-4 show how to adjust measurements to accommodate tile that is either 23½ or 24 inches long for two angles of slope in the speaker board, either

**5-12**   Project 3, showing internal construction of enclosures.

**5-13**   Project 4 before felt was glued onto speaker board.

**Table 5-3   Parts List for Project 4, High-Quality Small Speakers**

| Pieces | Dimensions | Function |
|---|---|---|
| 1 | 9 × 13″ ceramic flue tile | Enclosure shell |
| 2 | ¾ × 9 × 13″ plywood or particle board | Speaker board & back |
| 2 | ¾ × ¾ × 6⅝″ hardwood | Braces |
| 1 | ⅜ × 9 × 13″ plywood | Grille frame |
| 3 oz. | Polyester fill or Acousta-Stuf | Stuffing |
| 6 ft.² | Thin, foam-backed carpet | Wall lining |
| | **Components** | |
| 1 | 6½″ woofer | SEAS P17RC |
| 1 | 1″ inch soft dome tweeter | Polydax HD12X9D25 |
| 1 | 3.6 μF capacitor, Mylar, 100 V | C1 |
| 1 | 8.4 μF capacitor, Mylar, 100 V | C2 |
| 1 | 12 μF capacitor, NP electrolytic, 100 V | C3 |
| 1 | 4 Ω resistor, 15 W | R1 |
| 1 | 7.5 Ω resistor, 10 W | R2 |
| 1 | 0.3 mH inductor, air core | L1 |
| 1 | 0.4 mH inductor, air core | L2 |
| 1 | 0.32 mH inductor, air core | L3 |
| 1 | 8 Ω L-pad, 15 W | Tweeter control |
| 1 | Speaker terminal plate | |

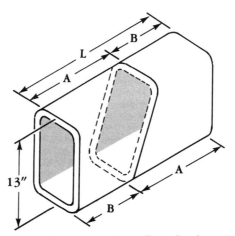

**5-14**   Cutting dimensions for Project 4. Refer to Table 5-4 for dimensions in inches.

**Table 5-4   Dimensions of Tile Parts Shown in Fig. 5-14**

| Slope | A | B |
|---|---|---|
| For L = 24: | | |
| 20° | 14⅜″ | 9⅝″ |
| 15° | 13¾ | 10¼″ |
| For L = 23½: | | |
| 20° | 14 | 9½ |
| 15° | 13½ | 10 |

15 or 20 degrees. The speakers shown in the photos were built and tested with a 20-degree slope.

When you buy your tile, remember to inspect it for hairline cracks. Some dealers are careless in handling it.

## Construction Procedure

Before cutting out the title enclosures, make sure you have protective goggles and a dust mask. Even so, the cutting should be done in a well-ventilated area because of the amount of tile dust thrown off by the saw.

You need a portable saw and a masonry blade. Masonry blades are not designed to be used at any angle other than 90 degrees, so the end cuts must be done carefully. Too much side pressure on the blade will cause it to shatter, a dangerous possibility. The end cuts at the top and bottom of the front end of the enclosures should be made first. Set the saw blade at the proper angle—20 degrees for the enclosures shown here—and adjust the depth for a very shallow cut, about 1/16 inch per pass. Check the depth carefully; don't try to saw deeper than 1/16 inch per pass with your saw. Make sure the tile is supported by a sturdy work table or the floor. After you make the end cuts, increase the saw depth by increments of 1/16 inch per pass until it cuts through the tile. Then turn the tile on its side to finish the job. At this point, remember to reset the saw blade to 90 degrees. Again, limit the depth of the cuts with each pass.

After cutting out the enclosure sections (FIG. 5-15), glue a lining of ¼-inch carpet over the interior of the tile. The right kind of carpet for this job is about ¼-inch thick, half of which is a layer of polyurethane foam backing. I found mine at K-Mart. The foam goes against the tile. Use silicone rubber sealant to install the carpet.

To brace the walls, cut a ¾-×-¾-inch piece of hardwood just

**5-15** A single piece of 9-X-13-X-24-inch flue tile, cut in half to make two enclosures.

long enough to be driven between the side walls (FIG. 5-16). Each end of this brace should be angled slightly to allow it to make tighter contact with each wall when it is not quite perpendicular to the walls. It will be slightly longer than the exact distance between the walls in your enclosures. You might have to cut and try several times to get it right, but it is worth the effort. The proper brace reduces any tendency for the side walls to vibrate.

Use the tile sections as patterns to mark the outlines of the speaker board and back before cutting them. I chose some surplus 13-ply Baltic birch plywood that I found in the discard pile of a furniture factory for the speaker board. Such plywood is far better than the typical building supply product. Even so, I braced the board with a ¾-X-¾-inch hardwood brace, glued and screwed behind the board in the space between the woofer and tweeter (FIG. 5-17).

I used ⅞-inch fiberboard for the back, but high grade plywood is acceptable. Mount the speaker terminals on the back before attaching it to the tile. The crossover network, diagrammed in FIG. 5-18, can either be installed inside or outside the back. If you put the network inside, you must cut a 1-inch hole in the back to allow access to the L-pad. Mount the L-pad on a small piece of tempered masonite, and

**5-16**   Tile sections lined with foam-backed carpeting and braced in the middle of each side. Ends of braces are angled slightly to make solid contact with walls when driven almost perpendicular.

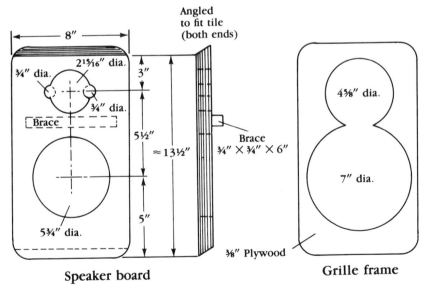

**Speaker board**                                    **Grille frame**

**5-17**   Dimensions for speaker board and grille frame for Project 4.

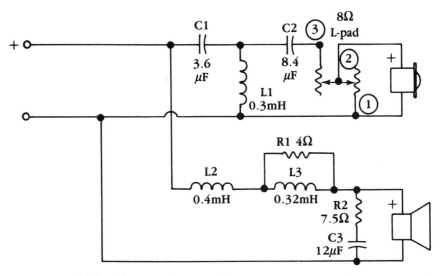

**5-18**  Schematic diagram of Project 4 crossover network.

install the masonite on the inside of the back panel with screws and glue so that the shaft of the pad will be accessible in the cutout.

Set the back on a flat surface with rails under it to protect the terminals. Run a bead of silicone rubber around the edge of the back and another around the rear edge of the tile. Set the tile on the back. The weight of the tile will make a good seal. Leave the enclosure in this position for a day. Then prop the enclosure up so that the front end is level. Fill it with about 3 or 4 ounces of loose polyester stuffing. Glue down the speaker board using weights to hold the board tightly against the tile.

After a day or more, use a rasp or coarse sandpaper to shape the edges of the speaker board and back to match the tile. You can cover the finished enclosures with contact covering, cloth, or even hardwood veneer. I painted mine flat black. To reduce diffraction, the face of the speaker board can be covered with foam or felt before installing a grille. Figure 5-17 shows the cutout dimensions for a grille frame that can be installed with brads or Velcro.

One of the questionable virtues of natural-sounding speakers such as these is that they reveal the enormous range in recording quality. There is no false bass here; it must be in the program. The bass is a bit lighter than that of larger speakers, but when there is true bass in the program, it comes out with a solidity that is surprising for small speakers. However, they will not rattle the windows. Their forte is clarity and definition, not sheer power.

P.S. Since the time that this project appeared in *Speaker Builder*, Ralph Gonzalez modeled this system with his computer Loudspeaker Modeling Program (LMP). He found that for optimum performance, the enclosure should be installed on a 15-inch stand. For further details of his study, including a derived frequency response graph, see the February 1989 issue of *Speaker Builder*, pages 57 and 58.

*Chapter* **6**

# Ported-Box Speaker Systems

$W$hen designing a closed-box speaker enclosure, you have a single major decision: the cubic volume of the box. For a ported-box system, you must choose not only the volume but also the resonance frequency of the port. Use the charts published in this chapter to tune the box to any suitable frequency, but first, you must find the proper frequency to match your woofer to its enclosure.

Before considering how to design a ported box, let's look at the various kinds and their characteristics.

## KINDS OF PORTED ENCLOSURES

Figure 6-1 shows three kinds of ported-box systems in general use today. The simplest type, FIG. 6-1A, is by far the most popular. Some ported boxes have an auxiliary diaphragm in the place of a port (FIG. 6-1B). Such a diaphragm is sometimes called an Auxiliary Bass Radiator (ABR) or a passive radiator (PR). These drone cones are like extra speakers without voice coils or magnets (although the diaphragm might be flat rather than cone shaped). The ideal ABR must have high compliance. It is tuned by adding mass to the diaphragm; the greater the mass, the lower the frequency.

The double-chamber reflex (FIG. 6-1C) is designed to load the driver over a wider band of frequencies than the single-chamber box. As the name suggests and the diagram shows, the double-chamber reflex has two compartments. The first compartment, behind the

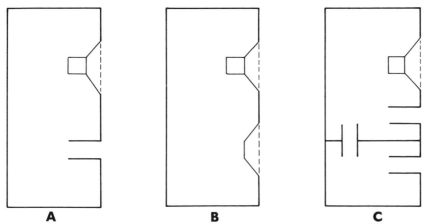

**6-1** Three kinds of reflex enclosures: (A) simple port, (B) auxiliary diaphragm, and (C) double-chamber reflex.

woofer, usually has twice the cubic volume of the second compartment. If ports of equal length are used for the two chambers and to connect them, the driver sees two resonance frequencies an octave apart. At the lower frequency, the driver behaves as if it were installed in a single-chamber box with a cubic volume equal to the total volume of the two compartments. The double frequency loading makes the double-chamber box somewhat more tolerant of driver variation than an ordinary ported box.

The tuned frequency of the box depends on the cubic volume of the box and the kind of port (FIG. 6-2). Each of the changes, moving from left to right in the figure tunes the box to a higher frequency.

Figure 6-3 shows how cone movement might vary for a woofer installed in a closed box, a simple reflex, and a double-chamber reflex. The line for the closed box shows a slight peaking of excursion near resonance as would occur for a speaker with a $Q_{TC}$ of about 1. These are not measured curves but rather typical kinds of curve for each type of enclosure. The dip in the simple reflex curve is purposely offset from the lower frequency dip in the double-chamber curve for clarity.

The point to note in this graph is that the area between the closed-box curve and the double-chamber reflex curve is greater than the area between the closed-box curve and the simple reflex curve. This indicates that for a music program that consists of a wide range of bass frequencies, the overall cone excursion would be less for a double-chamber reflex than for a simple reflex. Except for very low frequencies, either kind of reflex shows lower cone movement than the closed-box speaker. The advantage of the closed box is that

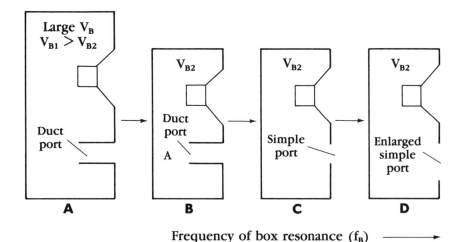

Frequency of box resonance ($f_B$) ⟶

**6-2** How box volume and porting affect tuning. Each change, going from left to right, raises the frequency of resonance.

it doesn't unload the cone at those low frequencies. This graph shows why a good infrasonic filter is almost a necessity for use with ported-box speakers.

## HOW BOX VOLUME AND PORTING AFFECT TUNING

As the driver cone moves against the air cushion in the box, a ported box behaves must like a closed box until the frequency range of the port is approached. The frequency of the port is determined by the mass of the air in the port vibrating against the compliance of the air in the box. When the driver vibrates at the port frequency, the air in the port "takes over" and vibrates more than the cone. The combination of vibrating air and enclosed air in the ported box form a resonator, properly called a *Helmholtz resonator* after the 19th-century German physicist who first described the behavior of acoustical resonators.

In FIG. 6-2A, the compliance of the enclosed air is greater than in B, C, or D because of the greater cubic volume in A. Reducing the cubic volume but maintaining the same port, as in B, raises the resonance frequency. In C, the volume is unchanged from B, but the mass of the air in the port is reduced by the removal of the duct behind the port. And in D, the simple port is enlarged, again raising the frequency of resonance.

Suppose you want to tune a box to a lower frequency. As you can see from FIG. 6-2, you can add a duct behind the port. Or, if there is already a duct, lengthen it. If the indicated duct length is too great for

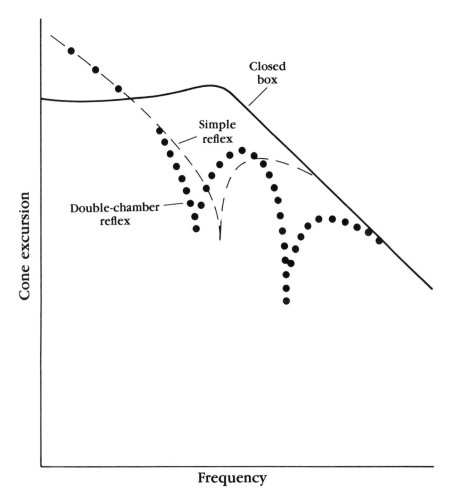

**Frequency**

**6-3**  Cone excursion versus frequency for a driver in a closed box, a simple reflex, and a double-chamber reflex.

the box depth, you can bend the duct into a L shape (FIG. 6-4). If you have a way of cutting perfect circles, plastic pipe can be used for ducts. To bend it, add a fitted elbow.

It is sometimes suggested that you can get the same performance from a small box as a large box if you use a longer duct on the small box. Such statements confuse tuning with performance. A small box can outperform a larger box only if the smaller box is a better match for the driver's Q and compliance.

One of the peculiarities of a reflex enclosure is that port radiation does not vary with the size of the port. A port can be so small that

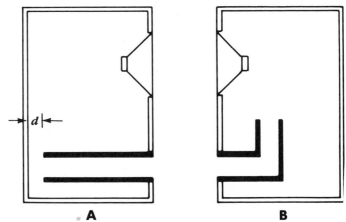

**6-4** How to install a long duct. (A) Incorrect if *d* is less than 3 inches. (B) Correct — effective length of this duct is the distance through the center of the duct.

the high air velocity in the port can produce unmusical noise and, if too small, it can impede the reverse air flow. The latter situation, allowing air to flow in one direction farther than in the other, is called *rectification*, which is an action similar to that of a rectifier on an alternating current. It can cause increased distortion.

One of the advantages claimed for the drone cone is that a small box can be tuned to a low frequency without resorting to very small ports or long ducts. This is true, but for a drone to equal a vent in output, it must be extremely compliant and probably should have a larger piston area than that of the driver.

## HOW BOX TUNING AFFECTS RESPONSE

In the days before Thiele/Small alignments, it was considered correct to tune every bass reflex enclosure to the free air resonance of the driver. Now we know why that didn't guarantee good performance. Drivers differ greatly in their degree of damping.

If a box is tuned too low for a given driver/box combination, the bass response will droop at a higher-than-desirable frequency. If tuned too high, there will be a bump in the response curve. Briefly, drivers that are highly damped, with Qs below about 0.38, usually should be installed in boxes that are tuned to a frequency above the free-air resonance. Drivers with Qs above that figure should be installed in boxes that are tuned to a frequency below the free-air resonance.

## HOW BOX VOLUME AFFECTS RESPONSE

Changing the box volume affects a ported speaker much the same as for a closed-box speaker. High-Q speakers need large boxes; low-Q speakers can be put in small enclosures.

What is a large box or a small box? That depends on the compliance of the speaker that is reported as "$V_{AS}$." $V_{AS}$ is the air volume whose equivalent compliance *for that driver* is equal to the driver's compliance. A box that looks large to you might "look" small to the driver. A 15-inch woofer, for example, can easily have a $V_{AS}$ of 20 to 30 cubic feet. If you installed such a woofer in a box of 5 cubic feet, the box would look rather large, but it would be acoustically small unless the woofer has a very low Q.

Keele's design formulas, as used in this book, allow an overvolume to account for a degree of unpreventable losses in a reflex system. Even the best reflex systems have leakage losses. For systems with well-built enclosures, the most serious losses probably occur in the driver itself, either through the suspension or the dust cover. Leakage losses are measured by the $Q_L$ of the system. A perfect system would have an infinitely high $Q_L$. In one test of a box with three different drivers, I found the system $Q_L$ to vary from 3 to 10. The driver that gave a $Q_L$ of 3 had a more open dust cover than the other drivers. A system with a low $Q_L$ requires a larger box for equal performance. Most charts assume a $Q_L$ of 7.

## THIELE PORTED-BOX DESIGN

A. N. Thiele's work has made the design process for ported boxes much more predictable. "Predictable" means within limits that are much narrower than those available before Thiele. Some design charts show figures carried out to several places to the right of the decimal point. Such precision is justified for the theory developed by Thiele, but achieving it with an acoustical system is something else.

If you have access to an IBM-compatible PC computer, you can use the program in Appendix B of this book. It generates a theoretical response curve for your woofer in any size box or with any tuning. With the computer, you can quickly explore how changes in box volume or tuning change response.

## KEELE'S CALCULATOR METHOD OF VENTED-BOX DESIGN

Again, as in designing a closed-box speaker, you need three kinds of Thiele data to design a ported-box system:

$f_s$— the driver's free-air resonance

$Q_{TS}$— the driver's total Q factor

$V_{AS}$— the driver's compliance stated as equivalent air volume

Using these values, you can determine the four critical aspects of a ported box. These are:

- Box volume ($V_B$)
- Box resonance frequency ($f_B$)
- System cutoff frequency, where response is down 3 dB ($f_3$)
- Hump or dip in passband

D. B. Keele, Jr. developed the pocket-calculator method of ported-box design used here. Two flowcharts explain the idea behind this program. Figure 6-5 shows the general procedure, and the

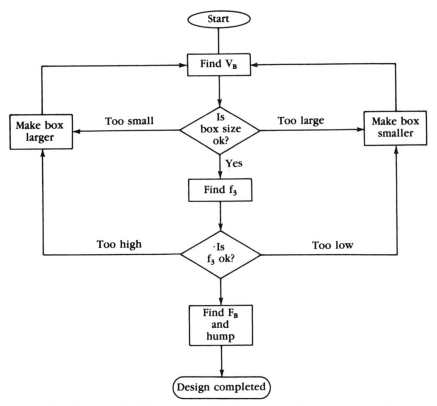

**6-5** Oversimplified flowchart of Keele's pocket calculator method of vented-box design.

specific formulas for the following procedure are in FIG. 6-6. If no
changes are necessary in the box size, use the first set of equations
(the left side of the chart) for $f_3$ and $f_B$; if any changes are made, use
the second set (the right side).

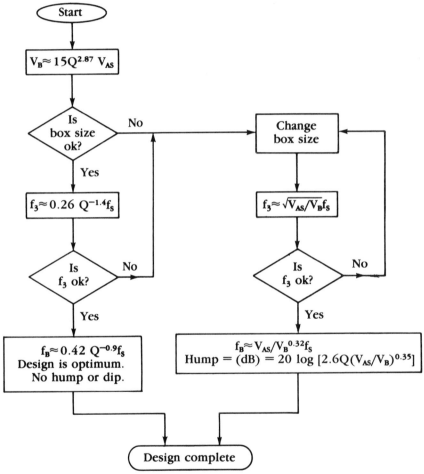

**6-6**  Simplified flowchart of Keele's pocket calculator method of vented-box design.

**PROCEDURE:**

$$V_B = 15 \; Q^{2.87} V_{AS}$$

If there is no change in box size, then

- $f_3 = 0.26 \; Q^{-1.4} f_s$
- $f_B = 0.42 \; Q^{-0.9} f_s$
- The design is optimum. There is no hump or dip.

If any change is made in box size:

- $f_3 = (\sqrt{V_{AS}/V_B}) f_s$
- $f_B = (V_{AS}/V_B)^{0.32} f_s$
- Hump (dB) $= 20 \; \log \; [2.6 \; Q \; (V_{AS}/V_B)^{0.35}]$

Refer to FIG. 6-6 throughout the following worked example. Figure 6-7 shows the response curves from the two design results obtained.

**EXAMPLE**

**Given:**

$$f_s = 21 \, Hz$$

$$Q_{TS} = 0.34$$

$$V_{AS} = 30 \, ft.^3$$

**Procedure:**

$$V_B = 15 \; (0.34)^{2.87}(30)$$

$$= 20.4 \; ft.^3$$

If no change in box size (optimum design):

$$f_3 = 0.26 \; (0.34)^{-1.4}(21)$$

$$= 24.7 \; Hz$$

$$f_B = 0.42 \; (0.34)^{-0.9}(21)$$

$$= 23.3 \; Hz$$

If box is too large, as a compromise try:

$$V_B = 10 \; ft.^3$$

$$f_3 = \sqrt{30/10} \; (21)$$

$$= 36.4 \; Hz$$

If $f_3$ is acceptable:

$$f_B = (30/10)^{0.32}(21)$$

$$= 29.8 \text{ Hz}$$

$$\text{Hump} = 20 \log [2.6\ (0.34)\ (30/10)^{0.35}]$$

$$= +2.3 \text{ dB}$$

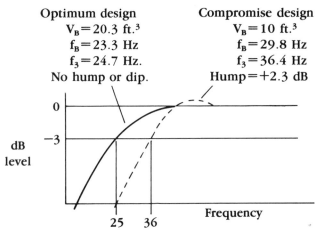

Optimum design
$V_B = 20.3$ ft.$^3$
$f_B = 23.3$ Hz
$f_3 = 24.7$ Hz.
No hump or dip.

Compromise design
$V_B = 10$ ft.$^3$
$f_B = 29.8$ Hz
$f_3 = 36.4$ Hz
Hump $= +2.3$ dB

**6-7**   Graph of response of driver in two boxes used in worked example of Keele's pocket calculator method of vented-box design.

To plan the right cubic volume, refer to Chapter 4. To tune the box, use the design tables in this chapter, TABLES 6-1 to 6-5. But before

**Table 6-1   Duct Length for Port with 3.14 in.² Area (2 in. Tube)**

| Vol. (ft.³) | 20 | 25 | 30 | 35 | 40 | 45 | 50 | 60 | 70 | 80 | 90 | 100 |
|---|---|---|---|---|---|---|---|---|---|---|---|---|
| | | | | | | | *Frequency (Hz)* | | | | | |
| 0.5 | | | | | | 7" | 5⅜" | 3¼" | 2" | 1¼" | | |
| 0.75 | | | | | 5⅝" | 4 | 3 | 1⅝ | ⅞ | | | |
| 1 | | | | 5½" | 3⅞ | 2¾ | 1⅞ | ⅞ | | | | |
| 1.25 | | | 6" | 4 | 2¾ | 1⅞ | 1¼ | | | | | |
| 1.5 | | 7⅝" | 4¾ | 3⅛ | 2 | 1⅜ | ¾ | | | | | |
| 1.75 | | 6¼ | 4 | 2½ | 1½ | 1 | | | | | | |
| 2 | | 5⅜ | 3¼ | 2 | 1¼ | | | | | | | |
| 2.5 | 7" | 4 | 2¼ | 1¼ | | | | | | | | |
| 3 | 5⅝ | 3 | 1⅝ | ⅞ | | | | | | | | |
| 3.5 | 4⅝ | 2½ | 1¼ | | | | | | | | | |
| 4 | 3⅞ | 2 | 1 | | | | | | | | | |
| 5 | 2¾ | | | | | | | | | | | |

**Table 6-2 Duct Length for Port with 7 in.² area (3 in. Tube)**

| *Vol.* | *Frequency (Hz)* | | | | | | | | | | | |
|---|---|---|---|---|---|---|---|---|---|---|---|---|
| *(ft.³)* | *20* | *25* | *30* | *35* | *40* | *45* | *50* | *60* | *70* | *80* | *90* | *100* |
| 0.5 | | | | | | | | 8⅜" | 5⅜" | 3¾" | 2½" | 1⅝" |
| 0.75 | | | | | | 10¼" | 7⅞" | 4¾ | 3 | 1¾ | ⅞ | |
| 1 | | USE | | | 9⅝" | 7⅛ | 5⅜ | 3 | 1⅝ | ¾ | | |
| 1.25 | | SMALLER | | | 7¼ | 5¼ | 3⅞ | 2 | ⅞ | | | |
| 1.5 | | PORT | | 8⅛" | 5¾ | 4 | 2⅞ | 1¼ | | | | |
| 1.75 | | | | 6⅝ | 4½ | 3⅛ | 2⅛ | ¾ | | | | |
| 2 | | | 8⅜" | 5½ | 3¾ | 2½ | 1⅝ | | | | | |
| 2.5 | | | 6¼ | 4 | 2½ | 1½ | ¾ | | | | | |
| 3 | | 8" | 4¾ | 3 | 1¾ | ⅞ | | | | | | |
| 3.5 | | 6½ | 3¾ | 2¼ | 1¼ | | | | | | | |
| 4 | 9⅝" | 5⅜ | 3 | 1⅝ | ¾ | | | | | USE | | |
| 5 | 7¼ | 3⅞ | 2 | ⅞ | | | | | | LARGER | | |
| 6 | 5¾ | 2⅞ | 1¼ | | | | | | | PORT | | |
| 7 | 4½ | 2⅛ | ¾ | | | | | | | | | |
| 8 | 3¾ | 1⅝ | | | | | | | | | | |
| 10 | 2¼ | ⅞ | | | | | | | | | | |
| 12 | 1¾ | | | | | | | | | | | |

**Table 6-3 Duct Length for Port with 14 in.² area ($3^3/_4 \times 3^3/_4$ in.)**

| *Vol.* | *Frequency (Hz)* | | | | | | | | | | | |
|---|---|---|---|---|---|---|---|---|---|---|---|---|
| *(ft.³)* | *20* | *25* | *30* | *35* | *40* | *45* | *50* | *60* | *70* | *80* | *90* | *100* |
| 0.5 | | | | | | | | | | 8¾" | 6¼" | 4½" |
| 0.75 | | | | | | | | | 7¼" | 4¾ | 3⅛" | 1⅛ |
| 1 | | | | | | | | 7⅜" | 4⅝" | 2⅛ | 1½ | |
| 1.25 | | USE | | | | | 9" | 5⅜" | 3 | 1⅝ | | |
| 1.5 | | SMALLER | | | | 9⅜" | 7 | 3⅞ | 2 | ¾ | | |
| 1.75 | | PORT | | | 10⅜" | 7⅝ | 5½ | 2⅞ | 1⅜ | | | |
| 2 | | | | | 8¾ | 6¼ | 4½ | 2⅛ | ¾ | | | |
| 2.5 | | | | 9¼" | 6⅜ | 4⅜ | 3 | 1⅛ | | | | |
| 3 | | | 11" | 7¼ | 4¾ | 3⅛ | 2 | | | | | |
| 3.5 | | | 8⅞ | 5⅝ | 3⅝ | 2½ | 1¼ | | | | | |
| 4 | | 12" | 7⅜ | 4⅝ | 2⅞ | 1½ | | | | | | |
| 5 | | 9 | 5¼ | 3 | 1⅝ | | | | | | | |
| 6 | | 7 | 3⅞ | 2 | ¾ | | | | | | USE | |
| 7 | 10⅜" | 5½ | 2⅞ | 1⅜ | | | | | | | LARGER | |
| 8 | 8¾ | 4½ | 2⅛ | ¾ | | | | | | | PORT | |
| 10 | 6⅜ | 3 | 1 | | | | | | | | | |
| 12 | 4¾ | 1⅞ | | | | | | | | | | |

### Table 6-4   Duct Length for Port with 25 in.² area (5 × 5 in.)

| Vol. (ft.³) | Frequency (Hz) | | | | | | | | | | | |
|---|---|---|---|---|---|---|---|---|---|---|---|---|
| | 20 | 25 | 30 | 35 | 40 | 45 | 50 | 60 | 70 | 80 | 90 | 100 |
| 0.5 | | | | | | | | | | | | |
| 0.75 | | | | | | | | | | 10" | 7" | 9⅜" |
| 1 | | | | | | | | | 9⅝" | 6⅜ | 4¼ | 2⅝" |
| 1.25 | | | USE | | | | | | 6⅞ | 4¼ | 2½ | 1¼ |
| 1.5 | | | SMALLER | | | | | 8⅜" | 5 | 2⅞ | 1⅜ | |
| 1.75 | | | PORT | | | | | | 6⅝ | 3¾ | 1⅞ | |
| 2 | | | | | | | 9⅜" | 5¼ | 2¾ | 1⅛ | | |
| 2.5 | | | | | | 9¼" | 6⅝ | 3⅜ | 1⅜ | | | |
| 3 | | | | | 9⅞" | 7 | 4⅞ | 2⅛ | | | | |
| 3.5 | | | | 11⅝" | 7⅞ | 5⅜ | 3⅝ | 1¼ | | | | |
| 4 | | | | 9⅝ | 6⅜ | 4¼ | 2⅝ | | | | | |
| 5 | | | 10⅞" | 6⅞ | 4⅜ | 2½ | 1¼ | | | | | |
| 6 | | | 8⅜ | 5 | 2⅞ | 1⅜ | | | | USE | | |
| 7 | | 11¼" | 6⅝ | 3¾ | 1⅞ | | | | | LARGER | | |
| 8 | | 9⅜ | 5¼ | 2¾ | 1⅛ | | | | | PORT | | |
| 10 | 12¾" | 6¾ | 3⅜ | 1⅜ | | | | | | | | |
| 12 | 10 | 4⅞ | 2 | | | | | | | | | |

### Table 6-5   Duct Length for Port with 49 in.² area (7 × 7 in.)

| Vol. (ft.³) | Frequency (Hz) | | | | | | | | | | | |
|---|---|---|---|---|---|---|---|---|---|---|---|---|
| | 20 | 25 | 30 | 35 | 40 | 45 | 50 | 60 | 70 | 80 | 90 | 100 |
| 0.5 | | | | | | | | | | | | |
| 0.75 | | | | | | | | | | | | 11⅞" |
| 1 | | | | | | | | | | 10½" | 7⅜ | |
| 1.25 | | | | | | | | | 10¾" | 7¼ | 4¾ | |
| 1.5 | | | | | | | | | 8 | 5 | 3 | |
| 1.75 | | USE | | | | | | 9⅝ | 6 | 3½ | 1¾ | |
| 2 | | SMALLER | | | | | | 7¾ | 4½ | 2⅜ | ¾ | |
| 2.5 | | PORT | | | | | 8⅞" | 5 | 2½ | ¾ | | |
| 3 | | | | | | | 6½ | 3¼ | 1 | | | |
| 3.5 | | | | | | 9⅜" | 4¾ | 1⅞ | | | | |
| 4 | | | | | 10½" | 7½ | 3⅜ | 1 | | | | |
| 5 | | | | 10¾" | 7¼ | 4¾ | 1⅝ | | | | | |
| 6 | | | | 8 | 5 | 3 | | | | USE | | |
| 7 | | | 9⅝" | 6 | 3½ | 1¾ | | | | LARGER | | |
| 8 | | | 7⅝ | 4½ | 2⅜ | ¾ | | | | PORT | | |
| 10 | | 8⅞" | 5 | 2½ | ¾ | | | | | | | |
| 12 | 11⅞" | 6½ | 3¼ | 1⅛ | | | | | | | | |

using those charts, you must choose a minimum port size for your woofer. To use a rule of thumb that works fairly well, divide the effective piston area of your woofer, $S_D$, by 7.

For a closer approximation, use the peak displacement volume of the cone, $V_D$. If $V_D$ for your woofer is not stated, you can find it by multiplying the piston area, $S_D$, by the maximum linear displacement, $X_{max}$.

Here is an example. For a woofer with an $S_D$ of 50 square inches, the rule of thumb would suggest a minimum port area of slightly more than 7 square inches, (50/7).

Or, use this formula (after R. Small):

$$S_V = 0.02 \ f_B \ V_D$$

where $S_V$ is in square inches, $f_B$ is in Hz, and $V_D$ is in cubic feet.

If the same woofer mentioned above has an $X_{max}$ of ⅛ inch (0.125 inch) and is to be installed in a box with a port frequency of 32 Hz, then:

$$V_D = 0.125 \ \text{in.} \times 50 \ \text{in.}^2$$

$$= 6.25 \ \text{in.}^3$$

And the minimum vent area is:

$$S_V = 0.02 \times 32 \times 6.25$$

$$= 4 \ \text{in.}^2$$

For a safety margin, double the minimum port area if possible. In this case, that would suggest a port area of 8 squares inches. A 3-inch tube, which has a cross-sectional area of about 7 square inches, should be acceptable. TABLE 6-2 shows the correct length for such a tube for various volumes and frequencies.

Note that you must sometimes estimate the length. If you want to use a 3-inch tube to tune a box with a cubic volume of 4½ cubic feet to 32 Hz, you won't find the exact length on the chart. Instead, find the length that tunes a box of 4 cubic feet to 30 Hz. That turns out to be 3 inches long. It would be a bit too long for tuning a 4½-cubic foot box to 32 Hz, but it is easier to cut off extra length than to add more. If the tube had been too long for the box, go to the previous table, TABLE 6-1. That table gives the length of 2-inch tube that would work. Or, if the length had been very short, you might want to go to the next larger size port tube.

## Keele's Calculator Equation for Vented Box Frequency Response

This is another useful mathematical model that goes quickly with a hand calculator.

### GENERAL FORM

**Given**:

Driver Parameters

- $f_s$ (Hz) — free-air resonance
- Q — total driver Q, sometimes labeled "$Q_{TS}$"
- $V_{AS}$ (ft.³) — compliance equivalent volume

Box Parameters:

- $V_B$ — net box volume
- $f_B$ — box frequency of resonance

**Compute**:

Constants

$A = f_B^2/f_s^2$

$B = A/Q + f_B/(7f_s)$

$C = 1 + A + f_B/(7f_sQ) + V_{AS}/V_B$

$D = 1/Q + f_B/(7f_s)$

Response in dB (normalized to 0 dB at high frequencies)

- At each frequency f (Hz), find: $f_n = f/f_s$
- Then substitute into the following equation:

$$\text{Response in dB} = 20 \log \frac{f_n{}^4}{\sqrt{(f_n{}^4 - Cf_n{}^2 + A)^2 + (Bf_n - Df_n{}^3)^2}}$$

**EXAMPLE:**

Given:

Driver Parameters

- $f_s = 33$ Hz
- $Q = 0.35$
- $V_{AS} = 17.9$ ft.$^3$

Box Parameters

- $V_B = 6.5$ ft.$^3$
- $f_B = 45$ Hz

**Compute:**

Find the response level at 45 Hz (down from midband).

Constants

$$A = (45/33)^2$$

$$= 1.86$$

$$B = (1.86/0.35) + 45/7(33)$$

$$= 5.31 + 0.19$$

$$= 5.51$$

$$C = 1 + 1.86 + 45/7(33)0.35 + 17.9/6.5$$

$$= 6.17$$

$$D = 1/0.35 + 45/7(33)$$

$$= 3.05$$

Response in dB where $f_n = f/33$:

For f   = 45 Hz:

$$f_n = 45/33$$

$$= 1.36$$

$$f_n^2 = (45/33)^2$$

$$= 1.86$$

$$f_n^3 = (45/33)^3$$

$$= 2.54$$

$$f_n^4 = (45/33)^4$$

$$= 3.46$$

Response dB =

$$20 \log \frac{3.46}{\sqrt{[3.46 - 6.17(1.86) + 1.86]^2 + [5.51(1.36) - 3.05(2.54)]^2}}$$

$$= 20 \log \frac{3.46}{\sqrt{37.9 + 0.064}}$$

$$= 20 \log 0.56$$

$$= -5.0 \text{ dB (computer program calculates } -4.9 \text{ dB)}$$

## Keele's Pocket Calculator Method of Computing Vent Dimensions

You can also use a calculator to find the right vent area/length ratio to tune your box to an exact frequency without using a chart. Once you find the ratio, you must juggle area versus length to obtain a vent that is neither too small in area nor too long for your speaker and box dimensions.

### GENERAL FORM

Given:

$V_D$—driver peak displacement volume

$V_B$—net box volume

$f_B$—box frequency of resonance

Compute:

Minimum vent area $S_{Vmin}$ (Based on large signal power output considerations to minimize vent noise) (after R. Small)

$$S_{Vmin} = 0.02 \, f_B \, V_D$$

Vent area length combinations ($S_V$, $L_V$)

- Compute $\alpha$ (vent area/vent effective length)

$$\alpha = V_B \ (2\pi f_B/c)^2 \approx 3.7 \times 10^{-4} V_B f_B^2$$

where $V_B$ is in ft.$^3$ and $f_B$ is in Hz

- Compute $L_V$

$$L_V = S_V/\alpha - 0.83\sqrt{S_V}$$

Start at $S_{V_{min}}$ and find $L_V$, step up with $2\,S_{V_{min}}$, $4S_{V_{min}}$, etc. Select $S_V$ and $L_V$ so that $L_V$ will not be too close to the back panel of the box. If possible, good values would be the area and length combination corresponding to:

$$S_V = 2\ S_{V_{min}}$$

**EXAMPLE**

**Given:**

- $V_D = 31$ in.$^3$
- $V_B = 6.5$ ft.$^3$
- $f_B = 45$ Hz

**Compute:**

$$S_{V_{min}} = 0.02\ (45)\ 31$$
$$= 27.9\ \text{in.}^2$$
$$\alpha = 3.7 \times 10^{-4}\ (6.5)\ (45)^2$$
$$= 4.9\ \text{in.}^2/\text{in.}$$
$$L_V = 27.9/4.9 - 0.83\ \sqrt{27.9}$$
$$= 1.3\ \text{in.}$$

**NOTE:** $L_V$ could come out negative; step up $S_V$ until $L_V$ is 0.75 inch (for ¾-inch material).

| Vent Length ($L_V$) | Vent Area ($S_V$) |
|---|---|
| 1.3 in. | 27.9 in.$^2$ $S_{V_{min}}$ |
| 5.2 in | 55.8 in.$^2$ $2\ S_{V_{min}}$ |
| 13.9 in. | 111.5 in.$^2$ $4\ S_{V_{min}}$ |

The best choice is the 5.2-inch vent length with 55.8-square-inch vent area.

D. B. Keele, Jr., who worked out the vented box design equations for the pocket calculator, says that this program works well with speakers that have a $Q_{TS}$ of 0.6 or below. When it is used with speakers of higher Q, the box volume will be huge and the tuning low. Such a system might have extended flat frequency response, but if driven higher than at very low sound levels, it can have high distortion.

## THE BOOM BOX ALIGNMENT

W. J. J. Hoge described a kind of alignment called a *Fourth-Order Boom Box* (BB$_4$). It has a peak near cutoff that gives the effect of good low-bass response. Bullock extended Hoge's alignment into a *Super Fourth-Order Boom Box* (SBB$_4$) for speakers with a lower $Q_{TS}$. The name *boom box* has bad connotations because it was formerly used for mistuned boxes. With the low-Q speakers of the SBB$_4$ alignment, there isn't much resemblance to those former boxes of ill fame (however, the BB$_4$ alignment sometimes earns its name). For speakers with a $Q_{TS}$ of 0.45 or lower, the boom is hardly noticeable in a properly tuned box.

An interesting feature of the Boom Box alignments is that the box is tuned to the free-air resonance of the woofer. The only design decision is to determine the correct cubic volume.

To design a Boom Box, use the chart in FIG. 6-8. Locate the value for the $Q_{TS}$ of your woofer on the left scale of the chart. Lay a ruler across the chart and read the value of line A where the ruler crosses it. You will find the correct value by following a vertical line from the point where the ruler touches A to the bottom of the chart. You can than find the volume for your enclosure by this formula:

$$V_B = (A)\ (Q_{TS})\ (V_{AS})$$

Here is an example. Suppose you have a woofer with a $Q_{TS}$ of 0.29 and a $V_{AS}$ of 3 cubic feet. The chart gives an A value of about 1.3. So:

$$V_B = (1.3)(0.29)(3\ \text{ft.}^3)$$
$$= 1.1\ \text{ft.}^3$$

To get the frequency of cutoff, use the curve B in FIG. 6-8. The formula for that is:

$$f_3 = (B)(f_s)$$

Going to the chart, you find that the ruler, in the same position as before, crosses the B line at about 1.9. If the woofer described above has a free-air resonance of 25 Hz, you can find the cutoff frequency by:

$$f_3 = (1.9)(25 \text{ Hz})$$

$$= 47.5 \text{ Hz}$$

In using the Boom Box alignment, follow the same rules as for the Keele data. That is, add extra volume to account for the space taken by components and braces.

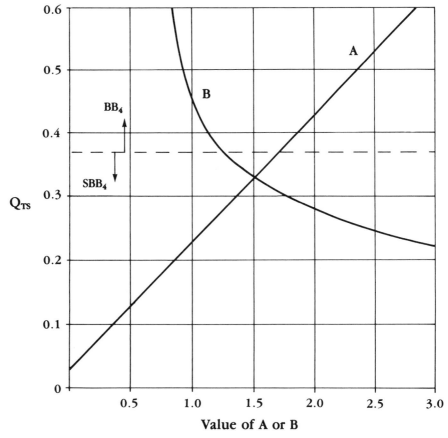

**6-8** Design chart for BB$_4$ and SBB$_4$ alignments. (See text for instructions.)

## HOW TO TUNE A PORTED BOX

If you have test equipment, you can tune a ported box to any desired frequency by adjusting the length of the port tube. The easiest way to tune a box is to use tubes, such as PVC pipe, and insert various lengths of pipe from outside the box while you check the tuning frequency. To do this, the tube must make an airtight fit with the hole cut for it. If you are using a rectangular duct, line one end of the duct with foam weatherstrip tape. For a rectangular port, the cutout in the speaker board should be the same as the internal dimensions of the duct. The total port length in that case will be the thickness of the panel plus the length of the tube.

When you tune from outside the box, there will be an almost unnoticeable shift in tuning when the duct is placed inside the box. The shift is upward in frequency but is not usually more than 1 or 2 Hz.

Sometimes it is convenient to build and tune an enclosure for a woofer you don't yet have. If you know the specifications of the woofer you plan to buy or already have on order, you can go ahead and build the box and tune it using Test 5, Method IV in Chapter 10.

When tuning the box with test equipment, start by making the port tube longer than the calculated length. Then adjust the tuning by cutting off the tube until you reach the desired frequency. To estimate how much length to cut off, use this formula (after Keele):

$$\Delta L_V = -\Delta f_B (2L_V/f_B)$$

where $\Delta L_V$ is the change in vent length, $\Delta f_B$ is the required change in frequency, $L_V$ is the length of vent when tested, and $f_B$ is the frequency of box as tested.

Notice that this formula gives a negative value for a positive $\Delta f_B$. To raise the box frequency, you must remove some of the length of the port. To lower the frequency of the box, $\Delta f_B$ would be a negative number, giving the solution a positive value. That would require a longer port tube.

Here is an example. Suppose you find that a 7½-inch port tube tunes a box to 30 Hz and you want to raise the tuning to 32 Hz. How much should you remove from the port tube? By the formula above:

$$L_V = -2 \text{ Hz } (15 \text{ in.}/30 \text{ Hz})$$

$$= 1 \text{ in.}$$

So you would cut an inch from the port tube, making it 6½ inches long.

When you tune from outside the enclosure and then later install the port tube inside the box, make sure the entrance to the tube is clear. Small tubes are especially subject to loss of efficiency if they are partially blocked. Even grille cloth can make a difference. To test the effect of your grille cloth, stretch a piece of it across the port while you drive the speaker to high output at $f_B$. If the cone shows a significant increase in movement when the material is placed over the port mouth, look for a more open material.

Too much damping material inside the box can make a change in tuning frequency. Typically, a reflex enclosure should have about an inch of fiberglass on the walls with an extra layer on the back behind the woofer.

## FINE TUNING BY EAR

The purpose of all this design theory is to produce a system that sounds good. If you go through the process and end up with a speaker that sounds either too weak in the bass or on the other hand boomy, you can always fine tune the box by ear.

If the bass is weak, cut some length off the tuning duct and listen again. If it begins to boom, you might have gone too far.

For a boomy system, try lengthening the duct. If that isn't possible, substitute a duct with a smaller cross-sectional area. In extreme cases, you might have to change woofers or build a larger box.

You might not get a system that fits the formulas by this kind of tuning, but you can often improve one that looks good on paper. After all, you are building a speaker system to listen to, not one to fit some formula.

## BURIED WOOFER REFLEX SYSTEMS

Woofers can be buried in a closed box speaker system, but you need a second woofer to radiate into the room. With a reflex system, it is possible to bury a woofer and allow the sound to emerge through ports. Figure 6-9A shows how one system works. When all the sound must go through a tuned cavity with a port, the system acts as a bandpass filter. This makes it useful only for subwoofer duty. One advantage for this kind of subwoofer is that you don't have to use any inductance in series with the woofer; the acoustical filter does the job of cutting off high-frequency response. With no reactances between

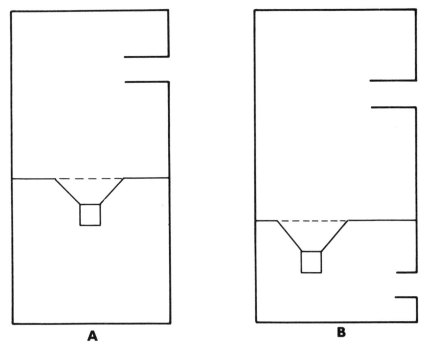

**6-9**   Two kinds of buried woofer reflex systems. Type B is patented by Bose.

the amplifier and the woofer, the full damping of the amplifier is applied to the woofer.

By choosing different alignments, you can alter the frequency response, transient response, and efficiency of this system.*

Here is an alignment for one woofer. It theoretically allows an efficiency equal to that of the same woofer in an ordinary box while maintaining uniform response and good damping. For those conditions, the value of Q' (the Q produced by the back volume) is set at 0.7. The value of the damping factor, S, is also 0.7. For this system, let's assume you want to design a box for a low-cost paper cone woofer, say the Pioneer B20. Its specifications are: $f_s = 35$ Hz, $Q_{TS} = 0.31$, and $V_{AS} = 1.9$ ft.$^3$.

To get the rear volume ($V_B$), follow the design method shown in Chapter 5. For a Q' of 0.7, that turns out to be 0.46 ft.$^3$. For the front volume ($V_F$), use this formula:

$$V_F = [(2) (S) (Q_{TS})]^2 V_{AS}$$

---

*To explore the many kinds of alignments, see *Speaker Builder* magazine, issue June 1988, page 29.

For the Pioneer woofer:

$$V_F = [(2)\ (0.7)\ (0.31)]^2\ 1.9\ \text{ft.}^3$$
$$= 0.36\ \text{ft.}^3$$

For the tuned frequency of $V_F$:

$$f_B = (Q')\ (F_s/Q_{TS})$$
$$= 0.7\ (35/0.31)$$
$$= 79\ \text{Hz}$$

From a design chart published in the magazine cited above, it appears that the bandwidth for this design is 81 Hz wide, from 50 to 131 Hz. To use this as a subwoofer, you would cross over to a satellite speaker at 131 Hz. The bandwidth is set by the choice of S and the frequency range by Q'. You can raise or lower the frequency as you like, but by lowering it, you lose efficiency.

Bose has patented another kind of double-chamber buried woofer system in which each chamber vents into the room (FIG. 6-9B). In a sense, such a double-chamber system consists of two separate enclosures. If the two enclosures were identical, the system wouldn't work because the sound from the back of the cone would be 180 degrees out of phase with that from the front of the cone. But if the frequencies of the two compartments are staggered an octave or so apart, the system can give useful output over a band of frequencies that includes the two resonance frequencies. The system can be used for a ratio of resonance frequencies of 1.5 to 3 and at volume ratios of from 2 to 4. While this system cuts off at the low end of its range at the usual reflex rate of 24 dB per octave, it has a high frequency cutoff at 12 dB per octave starting at about a quarter octave above the upper resonance. Again, this acoustical filter simplifies crossover design when using the system as a subwoofer.

Patented systems are protected from commercial use by anyone not licensed by the owner of the patent, but you can build such systems for your own use. To build a subwoofer like the one diagrammed in FIG. 6-9B, for example, you might set the larger compartment at three times the volume of the smaller one. To tune the compartments an octave apart, the port in the larger chamber should be about twice the length of the port in the smaller compartment. This assumes ports with equal cross-sectional areas. Treat each as a separate enclosure and use design charts to estimate cubic volumes and port dimensions. Expect to do considerable experimentation to get it right.

## DRONE CONE REFLEXES

Keith Johnson, who designs speakers for Precise Acoustic Laboratories, was quoted in *Speaker Builder* magazine on passive radiator designs. In essence, he said that of all the types of tuning, they are the most difficult. He mentioned the problems in various kinds of surround resonances, cone-drone interactions, and so on. There are also problems of specific cone-drone volume displacement ratios, compliance ratios, and the great possibility of a "boom box performance" even when care is used in the design.

Considering the problems of good PR design, it seems that this is an endeavor that is practical for the manufacturer who can spend considerable time and testing in order to work out the details for a long production run. For a do-it-yourself designer, there are more practical ways to go.

If you insist on doing a PR system, choose a drone cone with twice the piston area of the woofer and make sure you add enough mass for proper tuning.

## PROJECT 5: Ported-Box Speakers

This two-driver system is built into a box of Baltic birch plywood, but other materials can be used. If you want to use fiberboard, change the dimensions of the top and side pieces to make simple butt joints. The plans for the enclosure are shown in FIG. 6-10, and the parts are listed in TABLE 6-6. Note that the internal dimensions, as shown in the plans, are the measurements between plywood walls. The actual dimensions will be reduced by layers of asphalt roofing and thin carpet installed with Liquid Nails adhesive as described in Chapter 4. These lining materials go on the interior side walls, bottom, top, and back of the box. But before they are added, the box should be strengthened by an H-brace, as shown in the photograph in Chapter 4 (FIG. 4-4). Figure 2-2 in Chapter 2 shows a finished enclosure with drivers.

The crossover network (FIG. 6-11) is unusual because of the use of a Focal double-voice-coil woofer. The two voice coils make it possible to produce a three-way system with only two drivers. One voice coil of each woofer operates at low frequencies only, fed by a 4 mH inductor. The second voice coil works the midrange as well as the bass frequencies, served by a 0.7 mH coil. This double-voice-coil

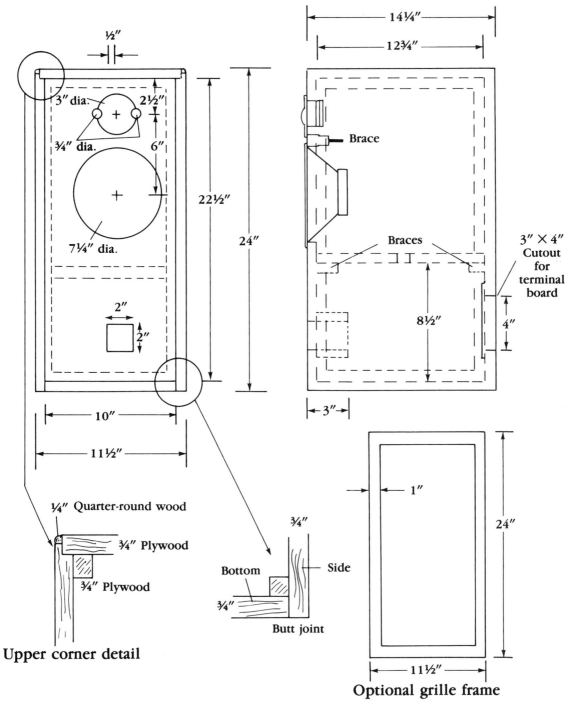

**6-10** Construction details for Project 5: Ported-Box Speakers.

**Table 6-6   Parts List for Project 5: Ported-Box Speakers**

| Pieces | Dimensions | Function |
|---|---|---|
| 2 | ¾ × 14¼ × 23¾" plywood | Sides |
| 1 | ¾ × 14¼ × 10" plywood | Bottom |
| 1 | ¾ × 14¼ × 11" plywood | Top |
| 2 | ¾ × 10 × 22½" plywood | Front & back |
| 1 | ¼ × 5 × 6" masonite | Terminal board |
| 1 | ⅜ × 11½ × 24" | Grille frame (optional) |
| 20 ft. | ¾ × ¾" hardwood | Cleats & braces |
| 6 ft.² | Asphalt roofing pieces | Wall liner |
| 6 ft.² | Thin, foam-backed carpet | Wall liner |
| 6 ft.² | Acoustical fiberglass | Wall liner |
| 2 | ¼ × 14¼" quarter-round wood | Exterior corners |
| 2 | ½ × 2¼ × 3" | Tuning duct |
| 2 | ½ × 2¼ × 2" | Tuning duct |
|   |   |   |
|   | **Components** |   |
| 1 | 8" woofer | Focal 8N401-DBE |
| 1 | 1" dome tweeter | Polydax HD 12X9 D25 |
| 1 | 6.8 uF capacitor, Mylar, 100 V | C1 |
| 1 | 8 Ω L-pad, 50 W | Tweeter control |
| 1 | 8 Ω resistor, 10 W | R1 |
| 1 | 1.5 mH inductor, air core | L1 |
| 1 | 18 uF capacitor, NP electrolytic 100 V | C2 |
| 1 | 0.7 mH inductor, air core | L2 |
| 1 | 4 mH inductor, iron core | L3 |
| 1 | Speaker terminal plate |   |

arrangement, with the first coil handling only the low bass, gives better low-frequency balance.

The Focal 8N401-DBE woofer is available at present but could soon be superceded by the 8N411-DBE. If so, the box design here can be tuned to the 8N411-DBE by reducing the length of the port by about ¾ inch.

This speaker should be installed on a stand, preferably one that tips the box back slightly. You can experiment with various elevations to find what works best in your listening room. For one possible base, see FIG. 3-4 in Chapter 3.

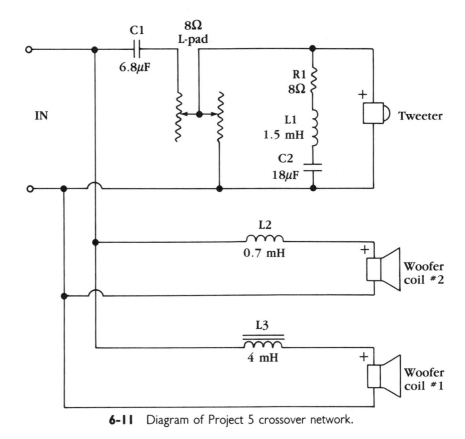

**6-11**   Diagram of Project 5 crossover network.

## PROJECT 6: Small Double-Chamber Reflex

This project demonstrates how to use Thiele data and Keele's formulas to design a double-chamber reflex for your woofer.

The double-chamber reflex described here (FIG. 6-12) was chosen to fill some existing spaces at each end of a small wall cabinet at a friend's cabin. The spaces were rather tall and shallow, which would have demanded a pipelike enclosure in a simple box. By dividing the longest dimension, the double-chamber reflex was perfect for our plans. A few calculations suggested a 6½-inch woofer.

Speaker Builder

**6-12**  Project 6: Small Double-Chamber Reflex system.

## Box Design

How do you design a double-chamber reflex? After looking at a design executed by George Augspurger many years ago, we came up with a four-step procedure:

**1** Choose the size woofer you want to use, or set the cubic volume and choose a suitable woofer. If you have test data for your woofer, make sure the total volume equals or exceeds the ideal vented box volume. If you have no test data, be generous.

**2** Set the lower frequency of resonance for the box. Remember, there will be two resonance frequencies in the completed enclosure. They will be spaced about an octave apart. In setting the lower resonance frequency, make it as low as or lower than the calculated $f_B$ for your woofer.

**3** Choose a convenient tube size to obtain a suitable vent area. Note that the area of a single tube must be multiplied by 2 to obtain the total vent area that interacts with room air (FIG. 6-13 and FIG. 6-14).

**6-13** Project 6 enclosure with PVC pipe ports.

**4** Use the total cubic volume and two times one tube area to find the duct length from a chart or by formula. This length will apply to all three tubes.

In designing the enclosure shown in FIG. 6-12, I began with a certain space to fill, about 1,500 cubic inches, and then went on to choose a woofer. At the time of this design (1985), the Peerless TP 165F had these specifications: $f_s$ = 53 Hz, $Q_{TS}$ = 0.42, and $V_{AS}$ = 15.8 liters, or 0.56 cubic foot. Running these values through Keele's calculator program for box design suggested an ideal box volume of 1,200 cubic inches tuned to 48.6 Hz with a theoretical cutoff frequency of 46.4 Hz. Using the total space available, 1,500 cubic inches, would produce a 25-percent overvolume. That is not a bad idea, because volume can be reduced easier than it can be enlarged.

It turned out that when the enclosures were built, my friend

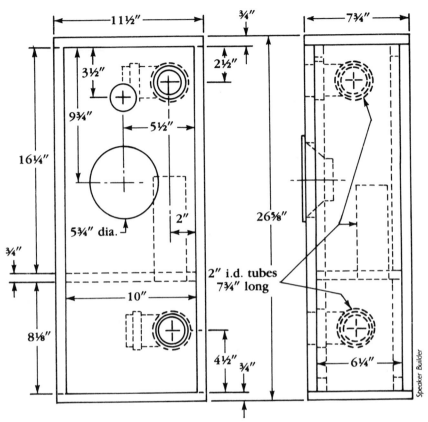

**6-14**   Construction plans for Project 6.

wanted to use them with some old Norelco woofers (Model 7065/ W8). The Norelco woofer had a lower Q, which would fit nicely into half the cubic volume tuned to 53 Hz with a 60 Hz cutoff. The chief cause of the mismatch was the low Q on the Norelco woofer. Taking a fresh look at the situation, we realized that with a Q of 0.38, the Norelco woofers could be used in the boxes we had built. It is easier to raise a Q than to lower it. In fact, Q is almost never what one might think it is because of the additive resistance in the speaker wiring and, particularly, the inductors in the crossover network. You can estimate the amount of resistance needed to raise the Q by this formula:

$$R = [(Q \text{ required}/Q \text{ measured})R_E] - R_E$$

For the Norelco, with a dc resistance ($R_E$) of 7 ohms, this turned out to be:

$$R = [(0.38/0.3)(7)] - 7$$

$$= 1.87 \text{ ohms}$$

When using this formula, remember to subtract any resistance in speaker wiring or crossover components from the theoretical value. In this system, we used less than the calculated resistance, preferring to err in that direction.

With a Q of 0.38, the Norelco would need a box tuned to 43 Hz, while the Peerless required a tuning of 48.6 Hz. Since there was a possibility that the old woofers might have to be replaced, we chose 45 Hz as a design aim. From the start, we had hoped to tune below 50 Hz to obtain loading in that frequency range. Checking out available kinds of vent tubes, we chose 2-inch PVC pipe. This has a cross-sectional area of 3.14 square inches, so the total port area to consider in calculating low-frequency tuning is 6.3 square inches. Plugging these values for port area and box volume into Keele's calculator program, we found that the vent tubes should be 7.6 inches long to tune the enclosure to 45 Hz. With an internal depth of only 6¼ inches, it was obvious that the two external tubes must be bent if they were to be mounted on the speaker board. When we bought the pipe, we bought four elbows of matching diameter.

**6-15**   PVC pipe details for two external vents of Project 6.

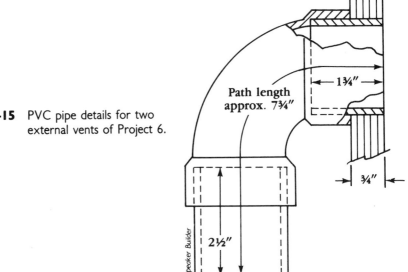

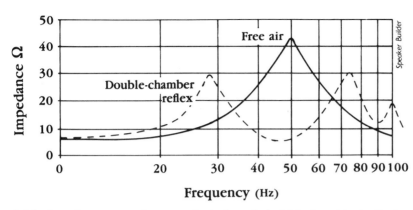

**6-16**   Impedance curve of one woofer in free air (solid line) and in the Project 6 double-chamber reflex (dashed line).

We estimated the length of the pipe/elbow combinations. As shown in FIG. 6-15, the 1¾-inch length, at least for the elbows we got, allowed 1 inch of pipe to fit into the elbow, leaving a ¾-inch projection to feed through the speaker board. This puts the shoulder of the elbow against the rear of the board, ensuring a better seal between pipe and board.

We glued the short sections into the elbows with a multipurpose PVC pipe cement. Then we stuck the pipes into the boards with silicone sealer. As the impedance curve in FIG. 6-16 shows, the lower box resonance occurred at about 46 Hz.

Although this project was described here mainly to show how

**Table 6-7   Parts List for Project 6: Small Double-Chamber Reflex**

| Pieces | Dimensions | Function |
|---|---|---|
| 2 | ¾ × 7¾ × 26⅝ | Sides |
| 2 | ¾ × 7¾ × 11½" | Top & bottom |
| 2 | ¾ × 10 × 25⅛" | Front & back |
| 1 | ¾ × 6¼ × 10" | Partition |
| 2 | 2(I.D.) × 2½" PVC pipe | |
| 2 | 2  "    × 1¾" PVC pipe | Front tubes |
| 2 | 2  "    90° elbows, PVC pipe | |
| 1 | 2  "    × 7¾" PVC pipe | Partition tube |
| 3½ ft.² | Acoustical fiberglass | Damping, upper chamber |
| | | |
| | Components | |
| 1 | 6½" woofer *See text* | |
| 1 | 2" tweeter *See text* | |
| | Crossover components, as required by above | |

you could go about designing a double-chamber reflex for any woofer, it can be used "as is" for a variety of 6½-inch woofers. Here is how you might use the enclosure with a current driver, say the Peerless 1429 woofer. That driver has the following characteristics: $R_E$ = 6.0 ohms, $f_s$ = 44 Hz, $Q_{TS}$ = 0.33, and $V_{AS}$ = 26 liters. Keele's calculator program suggests an ideal volume of 16 liters, while the enclosures described here have internal volumes of about 24 liters. But note that the free-air resonance of this woofer is very close to the lower frequency of the double-chamber box. By raising the Q of the woofer and using a boom box alignment, it might work. Looking over the chart in Fig. 6-8 and doing a few calculations, you can see that with a Q of about 0.44, it will work out right. For that, a 2-ohm resistance should be placed in series with the driver. In a real system, the resistance should probably be less than 2 ohms. Some audiophiles don't like to adjust Qs upward by adding resistance, but it is one way to adapt a driver to a box. You should always experiment with reduced amounts of resistance and use the lowest value possible.

Almost any small tweeter that will complement a 6½-inch woofer can be used. The Polydax TW74A, used in other projects in this book, is one possibility. It could probably be used with the Peerless 1429 woofer and a crossover network similar to that specified for Project 4B.

If you build a stereo speaker system using these enclosures, don't forget to make the two boxes as mirror images of each other. The speaker offset should normally be toward the space between the two enclosures. TABLE 6-7 lists the enclosure parts you need. You can substitute particle board for plywood. If you are using particle board and butt joints, the sides should be shortened by 1½ inches. The lengths given in TABLE 6-7 are for mitered joints at the corners.

The double-chamber reflex is a viable alternative to the single-chamber reflex, particularly if you want an extreme shape such as a tower.

*Chapter* **7**

# Transmission-Line Speaker Systems

$S$ome audiophiles consider transmission lines better than ordinary box speakers. A. R. Bailey, one of the first to advocate transmission lines, reserved his harshest criticism for reflex speakers. He considered their 24-dB-per-octave cutoff rate to be too sharp. Such devices, he said, were inherently more subject to ringing than systems with gentle roll-off.

Other advocates of the line say that it provides one way to get bass response from a woofer down to its free-air frequency—and, they add, with better quality bass. They often say the worst flaw of the typical speaker system is "boxy" sound.

I. M. Fried, another lover of the transmission lines, condemns closed-box speakers for their "oil can effect." "CB" speakers, he says, sound boxy because of the pressure build-up on one side of the cone. There is no doubt that such a pressure increase can cause nonlinearity and inhibit dynamic range in the bass.

Critics of transmission lines ask why build a large, expensive enclosure when you can get good sound from a smaller, easier-to-build box? They also cite the lack of precision in TL theory regarding either line design or desirable woofer characteristics. And, they say, it is hard to avoid coloration in the frequency band just above the low-bass region.

Regardless of arguments, pro or con, a transmission line is one option for speaker builders.

## KINDS OF TRANSMISSION LINES

There are many variations on the transmission line theme. Figure 7-1 shows some common kinds of lines and similar types of enclosures. Figure 7-1A is the traditional labyrinth. Labyrinths have a constant cross-sectional area for the tunnel behind the driver all the way to the port. The length of the line is equal to a quarter wavelength at the resonance frequency of the driver. At that frequency, it produces maximum damping on the driver cone. The walls of the labyrinth are lined, but the tunnel is more or less open.

When transmission lines were introduced, it was generally assumed that they are subject to the same rules as labyrinths. A few experimenters, notably John Cockroft, have brought this assumption into question.

Figures 7-1B and 7-1C show typical transmission lines. Instead of merely lining the walls, these enclosures are filled with damping material. In FIG. 7-1B, the line is almost constant in cross-sectional area with some enlargement near the driver. The port in that line, like the labyrinth, has the same area as the line. The line in FIG. 7-1C has a constant taper with a port no larger than the smallest area of the line. The taper suppresses line resonances.

Figure 7-1D shows a variant design—the tapered pipe. It can be considered a stopped pipe. Note that the taper on it goes the opposite direction to that of the tapered transmission line—more like a horn except that the port is considerably smaller in area than the largest section of the pipe. It is different from a transmission line in another significant way: the driver is installed near the midpoint of the line instead of at one end. Unless it is stuffed, it is a highly resonant device.

There are many other possible ways to make a transmission line. For example, some have the woofer at the bottom of the enclosure. In all kinds, the port is at one end, and the woofer is at the other. The path of the sound is such that its length is a significant fraction of the wavelength at low frequencies. This length, which is increased by stuffing and bends in the tube, produces a phase shift in the back wave of the woofer. This phase shift changes the out-of-phase back wave that would cancel the wave from the front of the driver into a reinforcement at low frequencies.

## DRIVERS FOR TRANSMISSION LINES

As mentioned earlier, there has been no great degree of mathematical precision in selecting drivers or in designing lines. It is generally

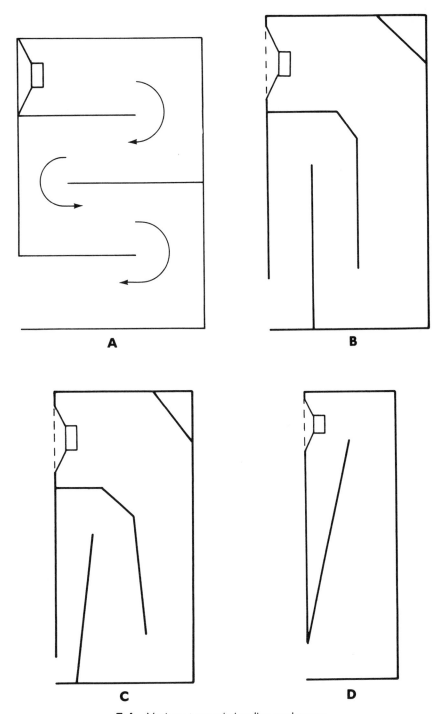

**7-1**   Various transmission line enclosures.

agreed that a low-Q driver is desirable. Any woofer that is designed to work well in a reflex enclosure should be a suitable choice for transmission line use. The line shares one characteristic with the ported box in that it tends to unload the driver at very low frequencies. Because of this, the power-handling ability of a woofer will be less in a transmission line than in a closed box. One way to obtain sufficient power-handling ability is to use a large woofer.

The minimum cross-sectional area of the line should be no less than the cone area of the driver. Add to that the fact that you need a long line (72 inches is a typical minimum), and you will have a very large enclosure, depending on the size of woofer you choose.

## STUFFED TRANSMISSION LINES

The stuffed line must be filled with damping material to suppress resonances and delay the travel of sound through the line. What kind of damping material? Almost every kind of material available has been used in transmission lines, but the favorite of purists seems to be long fiber wool. It has some disadvantages. First, it is expensive. For a one-time purchase that is no great matter, but it might not be a one-time purchase. Wool attracts moths and tends to settle. People who use it suggest washing and treating it with moth repellent. The second choice for many builders is polyester fill such as the kind sold in discount stores. It is satisfactory if carefully installed. Finally, Acousta-Stuf has sixteen bends per inch in its rather long fibers, which seems to make it perform better than ordinary fill materials. And there are no moth problems.

Different drivers and enclosures require different packing densities of stuffing for optimum performance. As an estimate, a packing density of about 8 ounces per cubic foot should be about right for a typical transmission line. But some lines are not typical. For description of those, see the next section on "Cockroft's Assumptions." Too much stuffing can overdamp the driver.

One problem peculiar to lines is how to suspend the stuffing so it won't settle. Some early transmission lines had dowels spaced throughout it to hold the damping material. Now more have the damping material suspended by wrapping it in nylon netting. The netting is stiff enough to hold the stuffing without collapsing. To use the netting, cut it in a piece somewhat wider than the length of a section of the line to be filled and long enough to roll up with the stuffing between the layers of netting. Distribute the damping material along the netting so that it will be arranged in an even packing density. If one end of the roll goes into a fatter section of the line, that end must have a greater amount of stuffing. If the roll doesn't look even, unroll it and repack until it is right.

Before installing the center fill, spray an adhesive on the inner walls and stick a layer of damping material there. Some experimenters say that sculptured foam is particularly good for lining the walls.

To properly stuff a line, you will probably have to do the job several times and compare the sound with each change. If you have facilities for running an impedance curve, try for the minimum impedance rise. But, again, you must use your ears to know if you have overdamped the line.

White stuffing will probably show in the port unless you dye it. Larry Sharp, who supplies Acousta-Stuf, says that you can use Rit dye to color it. Only dye the material that might be visible; don't dye the whole batch.

## COCKROFT'S ASSUMPTIONS ON TRANSMISSION LINE DESIGN

John Cockroft, a contributing editor of *Speaker Builder*, has recently offered some suggestions for designing lines for speakers with a higher-than-normal Q. He assumed that the standard line is 72 inches long with a driver that has a $Q_{TS}$ of 0.4. He further assumes that if you use a driver with a higher Q, you should make the line shorter, but fatter. Furthermore the packing should be made more dense in such a system. For Cockroft's "shortlines," there are no bends — just a straight tube. His shortlines are reminiscent of the Hartley "Boffle," which was very short and contained frames that held damping material with various sizes of openings in it. Cotton batting worked quite well in those enclosures.

If you have a woofer with a $Q_{TS}$ above 0.4, you can use the formulas here to design a shortline for it.

For cross-sectional area:

$$A = (Q_{TS}/0.4)^2 \times \text{cone area}$$

For line length:

$$L = 1/(Q_{TS}/0.4)^4 \times 72 \text{ in.}$$

For packing density:

$$D = (Q_{TS}/0.4)^2 \times 0.5 \text{ lb./ft.}^3$$

Note that if you try to use these formulas with a driver that has a

Q above 0.6, you get some weird figures. This suggests that you shouldn't try to use just any woofer in a transmission line.*

## PROJECT 7: Stuffed-Pipe Speakers

This stuffed, tapered pipe is an alternative to the bare-walled tapered pipes that were popular in Britain a generation ago. Voight was the first to use the "stopped" pipe, which is similar in action to an organ pipe. Similarly in 1949, Ralph West developed the Decca corner speaker. Both systems had relatively small drivers.

## Theory

The principle behind the pipe is shown in FIG. 7-2. The sketch in FIG. 7-2A shows relative pressure and velocity in a closed pipe at resonance. Pressure points occur at nodes; high-velocity points occur at loops. If you mount a speaker at the closed end of the pipe as in FIG. 7-2B, it will be loaded by the high pressure at that end, greatly increasing efficiency. However, two problems hinder that arrangement. First, the pipe's fundamental frequency is so strongly favored that low-frequency performance can sound like one-note bass. Another disadvantage is production of odd harmonics. The third harmonic is most serious (FIG. 7-2C). To correct it, place the driver at one-third the distance from the pipe's stopped end (FIG. 7-2D). At that point, the pressure is somewhat lower than at the closed end but is still high enough to provide good loading at the fundamental. At the third harmonic frequency, the drive point occurs at a loop instead of a node, reducing output.

Voight tapered the pipe to reduce the one-note effect and to spread the resonances over a band of frequencies (FIG. 7-2E). In later versions, the throat area was reduced to zero for smoother response, and the driver was installed at the midpoint of the line (FIG. 7-2F).

British builders used ⅜-inch plywood to construct their pipes. Thick walls were unnecessary, they said, because their enclosures were stiffest at the points of greatest pressure. Everyone agreed, however, that tight joints were imperative. One author said that a ⅟₁₆-inch gap in the driver mounting would reduce the output at 35 Hz by a factor of 4.

---

*For more on Cockroft's experiments, see *Speaker Builder* magazine of April 1988, p. 28.

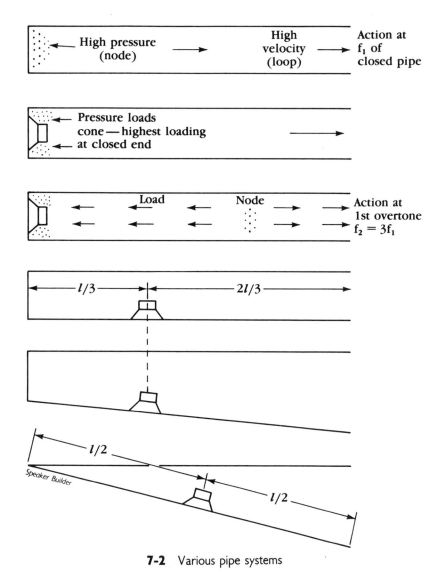

**7-2** Various pipe systems

## Pipe Design

To design a tapered pipe system, you must choose appropriate dimensions for pipe length, throat and mouth area, and drive point. The fundamental frequency is almost totally set by one parameter—the total line length. The formula given for a labyrinth or pipe length is:

$$L = (C/4f) - 1.7 \, r$$

where L is the length of pipe, C is the speed of sound in air, f is the pipe's fundamental frequency, and r is the pipe's radius or equivalent radius.

This formula, however, produces a longer-than-necessary pipe length. Opposing factors are at work here because the action of tapering the pipe tends to raise the fundamental frequency, while folding and choking it at the mouth lowers it. A bare pipe usually performs as though it were about 30 percent longer than its measured length. When you add damping material, it can seem even longer. As a rough estimate, expect the required length to be about 65 percent of the calculated length.

The pipe's cross-sectional area varies from zero or near zero to a maximum of about 2.5 times that of the driver's effective piston area. The pipe area should be about equal to the cone's area at the drive point, but ± 20 percent is acceptable. The port area is typically equal to the cone area, or about 0.4 that of the maximum area.

The taper rate of a pipe varies with the size of the driver. If the cross-sectional area starts at zero and goes to 2.5 times that of the cone, it will have to flare more rapidly for a pipe with a 6½-inch driver than that with a 4-inch driver unless the latter pipe is considerably shorter. Since line length is set by fundamental frequency, taper rate can be disregarded.

To design my pipes, I made rough sketches with estimated dimensions. I noticed that the drive point would be somewhat short of the midpoint from the throat to the mouth. You can only estimate the pipe length; the true acoustical length might be different. The formula for optimum drive point distance from the throat is:

$$d = L/(2 + \sqrt{A_t/A_m})$$

where d is the drive point distance from the throat, L is the pipe length, $A_t$ is the throat's cross-sectional area, and $A_m$ is the mouth's cross-sectional area.

The mouth is defined as the port area. The maximum area is that of the section just before the port. As you can see from the above formula, d can vary from ⅓ if the area of the throat is equal to that of the mouth, to ½ where the pipe tapers to zero throat area. Again, unless you use a straight pipe, you can only estimate the drive point's optimum location.

After doing some arithmetic, I found that by starting the throat at a point of 6 inches above the mouth and using an initial area of 2 square inches, the pipe would approximate the formula for the drive point. For a 4-inch woofer, the maximum area would be 24 square inches, which would be choked to 9.25 square inches at the mouth (FIG. 7-3).

## Test Results

When I ran an impedance curve of a 4-inch woofer in a bare pipe, FIG. 7-4, the curve looked similar to that of a woofer in a reflex. As I discovered later, the two systems behaved significantly differently in some tests. The lower peak was likely produced by cone loading from the air in the pipe, while the upper one was probably caused by the stiffness of the air directly behind the cone.

On listening tests, the bare pipe produced a hollow, ringing sound, particularly on male voices. I began to stuff the pipe, checking the sound after each addition of stuffing material. You can stuff a pipe until the impedance curve shows a single peak, like that of a closed-box speaker. The peak frequency, however, will be lower in the pipe than in the closed box. In my case, the peak occurred at 60 Hz. After stuffing and restuffing, I arrived at what seemed to be a good compromise between under- and overstuffing. The final amount in that first pipe was at the rate of about 8 ounces per cubic foot. The impedance curve for that condition is shown in FIG. 7-4 (dashed line). I used ordinary polyester fill in that pipe.

After the final stuffing, I compared the pipe speaker to performance with drivers of the same model in a closed box and a reflex enclosure. When listeners compared the three, the closed box speaker seemed to have the most limited range. Most initially thought the bass reflex had a bit more bass than the pipe, but after longer and more careful listening, that conviction waned. The pipe speaker performed better at the lowest frequencies. Frequency response tests showed the pipe to be clearly superior below 60 Hz.

After reviewing the experiments with a 4-inch woofer, I decided to work on a pipe designed for a 6½-inch woofer. It is the basis of this project.

## The Final Project

For the kind of sound you get per invested dollar, this pipe project is hard to beat. It requires careful adjustment of damping material for optimum performance, but otherwise it is easy to duplicate.

The project is built around a 6½ woofer and a low-cost piezo-electric tweeter (FIG. 7-5 and TABLE 7-1). Piezoelectric tweeters require no crossover network for protection because their impedance

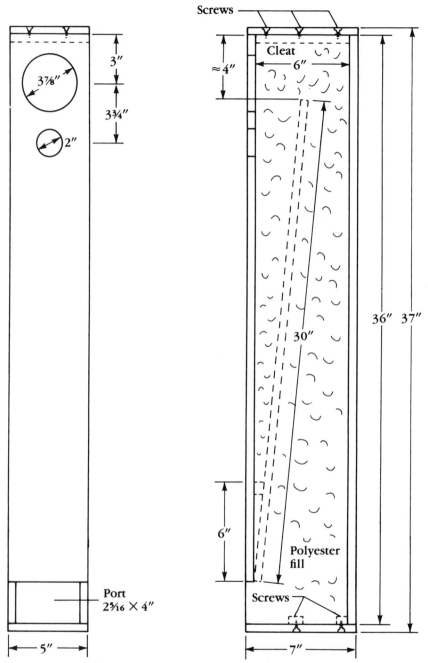

**7-3** Tapered pipe dimensions for 4-inch woofer.

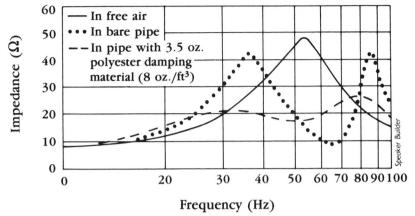

**7-4** Impedance curves of 4-inch woofer in free air and in pipes.

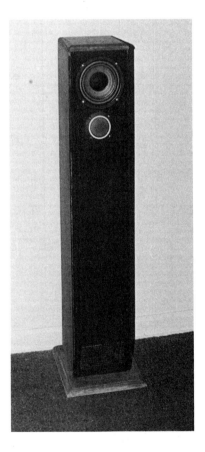

**7-5** Project 7: Stuffed-Pipe Speakers. The prototype of this project appeared in *Speaker Builder* magazine.

**Table 7-1   Parts List for Project 7: Stuffed-Pipe Speakers**

| Pieces | Dimensions | Function |
|---|---|---|
| 2 | ½ × 8 × 40¾″ particle board | Inner sides |
| 2 | ½ × 7½ × 40¾″ particle board | Inner front & back |
| 1 | ½ × 6½ × 30 particle board | Partition |
| 1 | ¾ × 9½ × 12¼ plywood | Bottom panel |
| 1 | ¾ × 9⅛ × 6½ plywood | Top panel |
| 6 ft. | ¾″ quarter-round wood | Top & bottom trim |
| 10 ft. | ¾ × ¾″ hardwood | Cleats & braces |
| 2 | ¼ × 7½ × 40¾″ | Outer front & back |
| 2 | ¼ × 9½ × 40¾″ | Outer sides |
| 1 | ½ × 8 × 40¾″ plywood | Optional grille frame |
| 6 oz. | Acousta-Stuf or polyester fill | Damping material |
| | **Components** | |
| 1 | 6½″ woofer | Pioneer polypropylene PC16 |
| 1 | Piezoelectric tweeter | Motorola 1039A |
| 1 | 4.7 uF capacitor, NP electrolytic 100 V | C1 |
| 1 | 8 Ω resistor, 10 W | R1 |
| 1 | 40 Ω resistor, 10 W | R2 |
| 1 | 0.5 mH inductor, air core | L1 |
| 1 | 8 uF capacitor, NP electrolytic 100 V | C2 |
| 1 | Speaker terminal | |

*Speaker Builder*

rises at low frequencies, blocking the bass. In this project, there is a filter in the tweeter circuit, but it is there to balance the tweeter's low end to better match that of the woofer (FIG. 7-6).

The only crossover elements in the woofer circuit (FIG. 7-6) are put there to form a notch filter. It smoothes the woofer's high-end response. Working together, the woofer and tweeter give you a lot of sound for the money, or, you might say, for peanuts.

What kind of sound? The first thing most people notice about this system is the bass response, unusual for a 6½-inch woofer. The next aspect they hear is the spatial depth, greater than in most speakers. That might be a bonus of the upright column. You might be tempted to enlarge this project to accommodate an 8-inch woofer. If you do that, good luck. I tried it and produced a larger tower than had poorer bass response than the ones shown here. It undoubtedly can be done, but be warned that it could take considerable experimentation.

Other woofers can be substituted. If you use another model, you can probably eliminate the notch filter. You can also substitute a

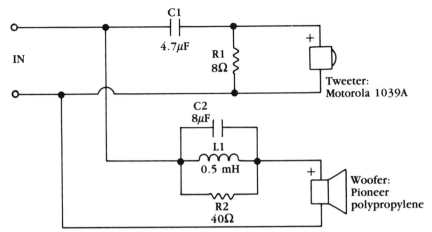

**7-6** Crossover network diagram for Project 7. If another woofer is used, eliminate C2, LI, and R2.

conventional tweeter, but that would require a complete crossover network, which would add considerably to the cost. The little piezo tweeter shown here occupies almost no space inside the pipe. A tweeter with a large magnet should probably be installed above the top.

## Construction Procedure

The enclosures shown in FIG. 7-5 are built with an unusual procedure that works quite well. It consists of building an interior enclosure with ½-inch particle board, then installing a veneer of ⅛- or ¼-inch hardwood plywood. You can glue the plywood outer skin on with wood contact cement, if you feel comfortable working with that adhesive. When you use it, your positioning of the parts must be right before they touch. As a novice wood worker, I feel better using Liquid Nails adhesive, a product that gives you time to make adjustments before it sets.

Here is a summary of the procedure.

**I** Cut out the parts for the inner enclosure from ½-inch particle board.

**2** To place the partition precisely, first set it in place on the inner surface of each side and mark an outline around it. Drill guide holes for nails down the centerline of that outline.

**3** After reversing the side, drive a nail through a hole at each end of the line of holes into the edge of the partition, just far enough to temporarily hold the partition in place. Use 3- or 4-penny finishing nails.

**4** Cut and shape the throat piece so that it fits into the space for it (FIG. 7-7) and is flush with the front edge of the sides. Then remove

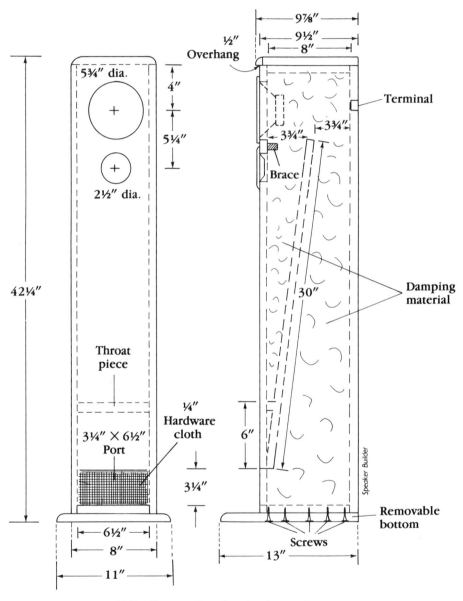

**7-7** Construction plans for Project 7.

the side from the partition. Glue and screw, or nail, the throat piece to the partition. Spread glue on the matching surfaces of the partition and sides. The prenailed holes ensure that the parts don't slip out of position as you drive in the remaining nails through the sides into the edges of the partition.

**5** After the glue sets, caulk the joints with silicone rubber sealer. Install the front and back panels with silicone rubber sealer as the adhesive. Nail the joints at close intervals, putting the nails about 3 inches apart. The use of the sealer, instead of conventional glue, prevents air leaks on the joints you can't reach for internal caulking.

**6** Install cleats inside the pipe to hold screws for top and bottom pieces.

**7** Use foam weatherstrip tape gaskets for removable parts, such as the top and bottom panels. One way to have improved access to the interior of the pipe is to make a temporary top piece with screws to hold it down. Then after all experimentation with damping material is completed, add a final top piece with glue and finishing nails.

**8** Add a ⅛- or ¼-inch veneer of hardwood plywood to all external surfaces to stiffen the panels and improve the appearance of the enclosures. The measurements in the plans allow for ¼-inch plywood, but even ⅛-inch plywood appears to be adequate in thickness.

Figure 7-8 shows further details of the enclosure, including two methods for attaching a grille frame. The grille can be just long enough to cover the drivers if you finish the front panel. Another small grille panel could cover the port.

Mount the crossover parts inside the back panel, in the rear top corner. At that location they will occupy some of the extra volume and may help break up reflected waves. Don't let the resistors touch other parts, and separate the inductors by a couple of inches. Glue them down with silicone rubber or other adhesive.

For stuffing, I found polyester fill satisfactory, but my final choice was Acousta-Stuf. Before placing the stuffing in the pipe, I glued a few pieces of ½-inch sculptured foam, obtained from packing cases for electronic gear, to the walls behind the woofer. Keep foam away from wiring and crossover parts to reduce fire danger.

Before installing damping material, carefully separate, or fluff out, the fibers in it so that the packing density is uniform. This takes considerable patience but is worth doing. You can install the stuffing by rolling it into a spiral with nylon netting or you can spray 3M

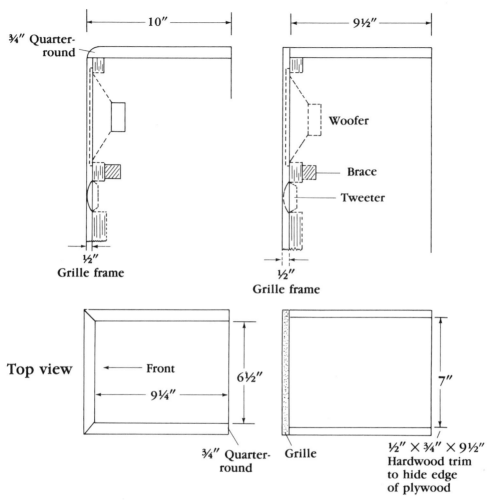

**7-8**  Details of two ways of installing a grille frame and grille cloth on Project 7. The prototype of this project appeared in *Speaker Builder* magazine.

adhesive on the walls to hold it in place. In areas of narrow cross section, settling is no problem.

In case you want to use the plans for a 4-inch woofer and small tweeter (FIG. 7-3), use drivers similar to those for Project 2. Or you can use a full-range 4-inch speaker and eliminate the tweeter.

## PROJECT 8: Transmission-Line Subwoofer

Here is a project for those who like large enclosures. Ideally it should be built of 1-inch fiberboard, but you can use ¾-inch material and cover it with plywood (FIG. 7-9).

**7-9** Project 8: Transmission-Line Subwoofer, in construction.

TABLE 7-2 shows how you can use the same plan for woofers of two sizes by simply changing the width of the enclosure. One possible application, install a double voice coil woofer and feed the bass from both channels to it. A pair of satellite speakers can handle the upper bass, midrange, and highs. To do that you should use a crossover network that limits the subwoofer to frequencies below 200 Hz. Another possibility would be to build a pair of these enclosures with midrange driver and tweeter in the upper front panels of each.

Figure 7-10 shows the dimensions and the plan for an easy-to-build enclosure. Figure 7-11 includes some refinements that require more careful work but will improve the performance of the system. To make the rounded corners shown there, you can use linoleum or plastic countertop material for a form and fill the space behind it with

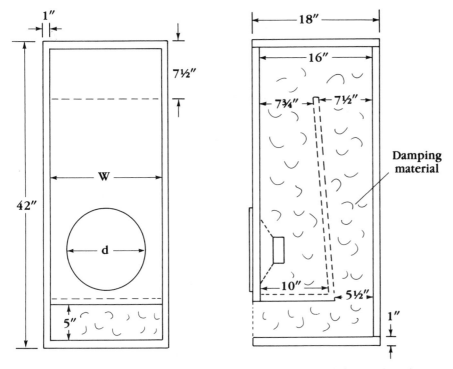

**7-10**  Construction plans for Project 8. Refer to Table 7-2 for width and speaker cutout dimensions.

concrete. To hold the concrete, first drive several screws into the walls of the enclosure where the concrete contacts them. Use large screws that are at least 1½ inches long. Drive them in about ½ inch so that a full inch of screw, including the head, will be embedded in the concrete.

**Table 7-2   Dimensions of Project 8 (Transmission-Line Subwoofer) Enclosures for Two Sizes of Woofers**

| Speaker size<br>*Advertised<br>diameter (in.)* | d<br>*Speaker<br>cut-out (in.)* | W<br>*Internal<br>width (in.)* | *Minimum<br>area (in.²)* |
|---|---|---|---|
| 10 | 9⅛ | 10 | 50 |
| 12 | 11 | 16 | 80 |

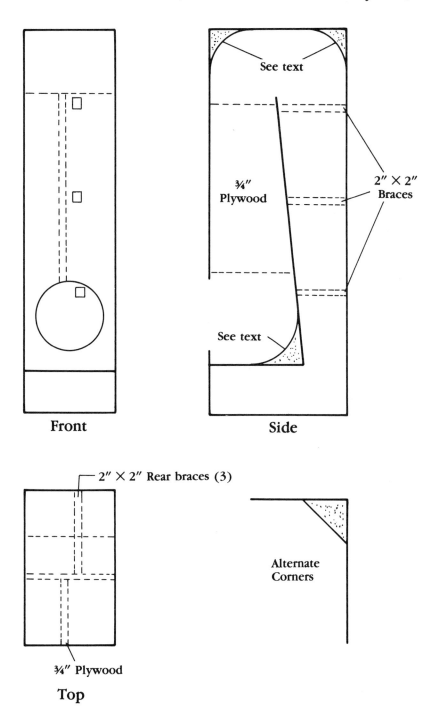

**7-11** Refinements for Project 8 that can improve performance.

   As an alternative method of treating the sharp corners, you can install wood forms for the concrete as in FIG. 7-11. If you use wood, treat the wood piece that serves as a form with motor oil before inserting the concrete. The motor oil will prevent the concrete from sticking to the wood so that the form can be removed. If you leave the wood, there is a chance that it will shrink away from the concrete and vibrate. After removing the wood, make up a mixture of cement and water and "wash" it over the surface of the concrete to make it more reflective.

   This rounding of the corners eliminates some of the undesirable reflections in a line. As a home constructor, you can add refinements such as these that cost you little more than time but improve quality.

# Chapter 8

# Crossover Networks

$M$ultiple-driver dynamic speaker systems need a crossover network to divide the frequency spectrum. The crossover can be simple, such as the one in Project 2, or it can be complex. There are arguments for and against both kinds.

## THE ROLE OF THE CROSSOVER NETWORK

The primary purpose of a crossover network is to feed highs to the tweeter and lows to the woofer. That is always one function, but there are others. In many speaker systems, the crossover network adjusts efficiencies and corrects for impedance variation. In some cases, it also corrects for frequency response unevenness. (Some of the special ways to correct speaker problems with a crossover network are discussed in Chapter 9.)

One of the most practical purposes of any crossover network is to protect small tweeters from overload at low frequencies. Without a filter to remove the bass, a tweeter would produce high distortion at best. Even worse, it would be damaged or destroyed. For this reason, some tweeters are operated with a network of a higher order (sharper cutoff) than the woofers.

To design a crossover network for your speakers, you must choose the rate of cutoff as well as the frequency of crossover. You can also choose between two general types of circuits, parallel or series. Even if you don't plan to design a network, it's useful to know how the various kinds work.

## KINDS OF NETWORKS

Crossover networks are named by the number of drivers in the system, by the sharpness of the cutoff action, and by the way the speakers are wired into the circuit. Two-way networks serve a woofer and a tweeter. The woofer is fed by a low-pass filter, the tweeter by a high-pass filter. If a midrange driver is added, you have a three-way system. The part of the network that feeds the midrange driver is called a bandpass filter because it passes a band of frequencies that are above those handled by the woofer but lower than those sent to the tweeter.

The simplest networks are called first-order networks. A two-way first-order network consists of a single element in each leg of the circuit. It produces a gentle cutoff slope that allows the woofer and tweeter to respond to a considerable band of frequencies beyond the theoretical crossover point. A first-order network, as diagrammed in FIG. 8-1, produces a cutoff slope of 6 dB per octave. First-order networks are simple to design and make. They have good phase characteristics in that the phase shift caused by each leg of the network is equal and opposite in phase. These differences theoretically cancel to produce a "minimum phase" crossover. It also adds less inductance in the woofer circuit, inductance that can reduce the damping on the driver. It also avoids other, even more pernicious, effects in some crossover circuits in which components interact with each other and with the amplifier.

One argument for simple crossover networks that is hard to deny is that some of the highly rated, and expensive, audiophile speaker systems use first-order networks. If sharper networks were clearly superior, as some advocates claim, one would expect to find them in all expensive speaker systems.

One designer who favors low-order networks suggests that the gradual shift from the limited dispersion characteristics of a large driver to the much wider dispersion of a small tweeter is more natural sounding than the sudden shift with a high-order crossover.

This gradualness of the first-order network causes adjacent drivers to operate in the same range over a considerable band of frequencies below and above the crossover point. This shared bandwidth can cause lobes of response (*lobing*) because of the phase shift produced by the distance between drivers (FIG. 8-2). Lobing occurs with higher order networks also but over a narrower bandwidth.

Second-order networks (FIG. 8-3) have two elements in each leg: a capacitor and an inductor. They produce cutoff rates of 12 dB per octave. The sharper cutoff gives extra protection to small tweeters

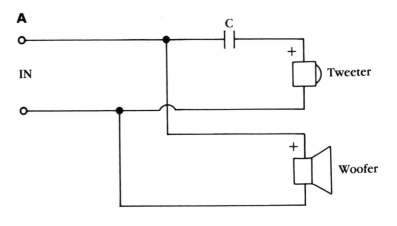

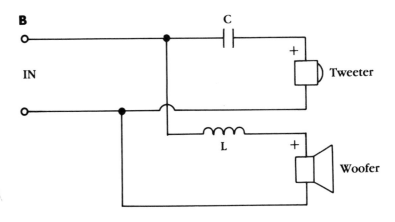

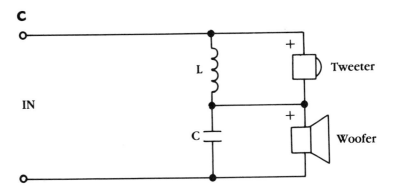

**8-1** Three kinds of first-order crossover networks: (A), high-pass filter only, (B), full first-order parallel, (C) and first-order series.

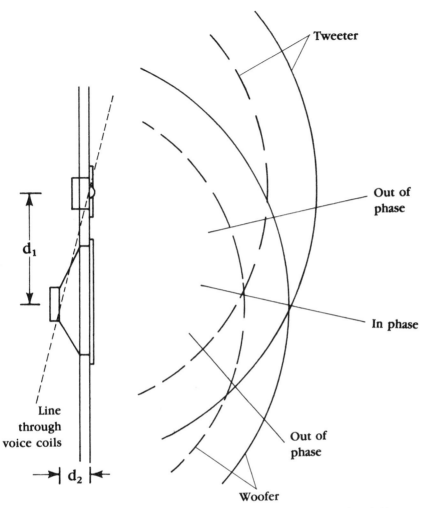

**8-2**   How the output from two drivers, displaced in space, can produce lobing because of phase differences where wave fronts overlap. This diagram assumes an equal signal to each driver.

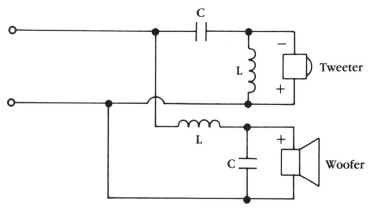

**8-3** A second-order parallel crossover network.

that are operated near the lower limits of their normal bandwidth. The advantages of a sharper cutoff are a more restricted band of shared duty and better protection for tweeters.

But there are also disadvantages. For one, the phase shift in a second-order network is 180 degrees, which means that the two drivers can be connected with like polarity at the expense of a null in the response curve at the crossover frequency. The normal reaction to this is to connect the woofer and the tweeter with negative polarity as in FIG. 8-3. This causes the high frequencies to be out of phase with the low frequencies. Whether this is significant is debatable.

The traditional second-order network is called a Butterworth network. When the drivers are connected in reverse polarity with a Butterworth second-order network, there is a 3 dB peak in the frequency response at the crossover frequency. With this network, you must choose between the peak or a null at that point. Because of this dilemma, another type of network, the Linkwitz-Riley, has increased in usage during the last few years. Considering the differences in direct frequency response, the L-R network appears to be superior. But in total power response, the situation is reversed: the Butterworth is uniform while the L-R has a dip. Briefly, the Butterworth might be a better choice for PA systems used in large auditoriums, while the L-R network would seem to be preferred for ordinary rooms.

For comparison, the design table, TABLE 8-1, also shows values for another second-order network, the Bessel. It is characterized by a slight peak at the crossover frequency, 1.25 dB, and a moderate dip in power response of 1.75 dB.

Various manufacturers of speaker systems use each of the net-

**Table 8-1   Multipliers to Convert First-Order Crossover Network Values for Second-, Third-, or Fourth-Order Networks**

| | First Order | Second Order | | | Third Order | Fourth Order |
|---|---|---|---|---|---|---|
| | | Butter-worth | L-R | Bessel | | |
| C1 | N | 0.71 | 0.50 | 0.57 | 0.67 | 0.53 |
| C2 | (From | 0.71 | 0.50 | 0.57 | 2.00 | 1.06 |
| C3 | chart | | | | 1.33 | 1.60 |
| C4 | or | | | | | 0.36 |
| | formula) | | | | | |
| L1 | | 1.41 | 2.00 | 1.74 | 0.75 | 0.63 |
| L2 | | 1.41 | 2.00 | 1.74 | 1.50 | 2.83 |
| L3 | | | | | 0.50 | 1.89 |
| L4 | | | | | | 0.94 |

works described above. The choice of a network probably has less effect on the total sound of a system than any one of several other design decisions. Lacking any special reason to choose one of the others, the L-R network is probably a safe bet.

Third-order networks (FIG. 8-4) have three elements in each leg and produce an 18-dB-per-octave cutoff rate. Finally, fourth-order networks (FIG. 8-5) have four elements in each leg and produce a 24-dB-per-octave cutoff.

The diagrams of two-way systems assume that every crossover will fit the "book" in a symmetrical arrangement of the same order for the woofer circuit as the tweeter circuit. In practical usage, there are many variations. For example, a first-order woofer circuit is often combined with a second- or third-order tweeter circuit. Such a combination is practical, because the higher order gives extra protection for the tweeter.

Two kinds of three-way networks are shown in FIG. 8-6. Those networks are both first-order. Because the crossover frequencies can be set at some distance from the response band limits of the woofer and tweeter in a three-way system, you can usually get satisfactory performance with a network of more gentle cutoff slope.

In addition to their order, networks can be identified by the kind of wiring circuit—series or parallel. Most engineers prefer the parallel network because they can treat the wiring of each driver as a separate circuit. This is particularly useful if you want to use bi-wiring

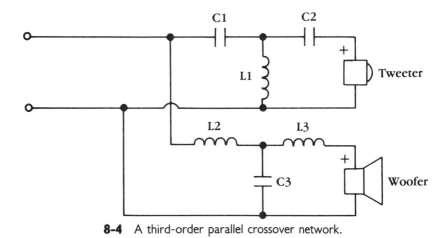

**8-4** A third-order parallel crossover network.

for the system. All of the networks in FIGS. 8-1, 8-3, 8-4, 8-5 and 8-6 are parallel networks except for the ones shown in FIG. 8-1C and FIG. 8-6B.

## HOW TO DESIGN CROSSOVER NETWORKS

There seems to be a trend toward two-way rather than three-way speaker systems. It has become apparent that the crossover point poses a potential for problems of unevenness in response or unnatural phase conditions. Considering the problems of getting the crossover right, it seems only prudent to favor two-way speakers except in the largest systems. This is especially true if you plan to design your

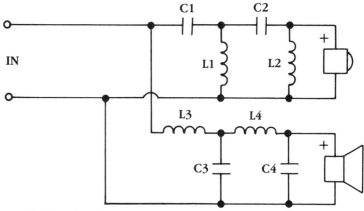

**8-5** A fourth-order parallel crossover network.

**A**

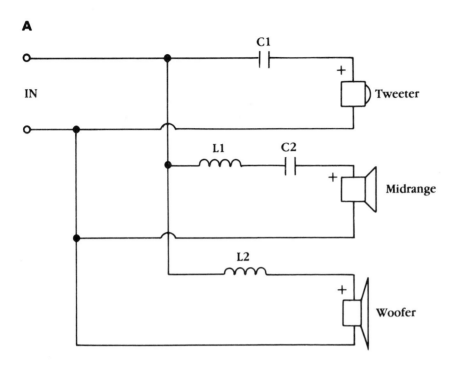

**B**

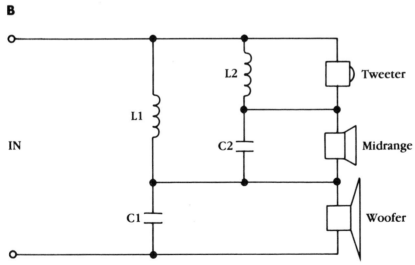

**8-6** Two kinds of three-way, first-order crossover networks: (A) parallel, and (B) series.

own system. By keeping it simple, you improve your chances of success.

## Parallel Networks

The simplest kind of crossover is a high-pass filter put in series with the tweeter to block the bass, as shown in FIG. 8-1A. A capacitor has more reactance to low frequencies than to highs, so a single capacitor reduces the signal to the tweeter at a rate of 6 dB per octave below the crossover frequency. The *crossover frequency* is defined as the frequency where the response is down 3 dB. This occurs at the frequency where the impedance of the capacitor equals the impedance of the tweeter. To move the crossover point up to a higher frequency, you would choose a smaller capacitor.

Capacitors are measured in farads, but the values useful for crossover networks are stated in microfarads ($\mu$F). The formula for capacitance in a high-pass filter is:

$$C = 1/(2 \, \pi \, f X_c)$$

where C is the value of the needed capacitance, f is the desired crossover frequency, and $X_c$ is the necessary capacitive reactance at the crossover frequency. As stated earlier, the reactance should equal the impedance of the driver.

The formula above gives the capacitance in farads, so multiply it by 1,000,000 to get the value in microfarads. If you prefer to get the answer in microfarads, use this formula:

$$C = 159000/(f R_T)$$

where f is the crossover frequency and $R_T$ is the impedance of the tweeter.

You can use the formula for tweeters with any value of impedance and for any crossover frequency. For another way to find the correct value, go to the chart in FIG. 8-7. For example, if you have an 8-ohm tweeter and you want to choose a capacitor that gives a reactance of 8 ohms at 4000 Hz, follow the vertical line up from 4000 Hz to the horizontal line for 8-ohm speakers. Then move diagonally to the upper left on the dashed line to find the correct value of capacitance. In this case it is 5 $\mu$F.

If you don't have a capacitor of the right value, you can wire two or more in parallel. Remember that capacitors in parallel have a total capacitance equal to the sum of the individual capacitors.

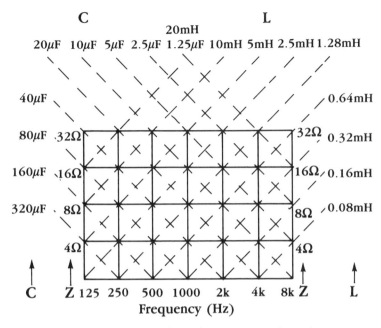

**8-7**   Crossover network design chart. Approximate values of inductance and capacitance for speakers with impedance ratings from 4 to 32 ohms.

To complete a first-order network, place a coil in series with the woofer (FIG. 8-1B). A coil has the opposite frequency characteristic to that of a capacitor. The coil's reactance to an alternating current increases with the frequency of the current.

Even a few years ago, it was hard to find coils with a good selection of inductance values. Now that has changed, and you can find almost any desired inductance.

Inductances are measured in henries, but the values useful for crossover networks are stated in millihenries (mH). The formula for inductance for a first-order crossover is:

$$L = X_L / (2\pi f)$$

where L is the value of inductance, $X_L$ is the inductive reactance, and f is the desired crossover frequency. The inductive reactance should equal the impedance of the woofer.

The formula above gives the inductance in henries. So multiply by 1000 to get the value in millihenries. To get the answer in mH, use this formula:

$$L = 159R_w/f$$

where $R_w$ is the impedance of the woofer and f is the crossover frequency.

Again, if you prefer, you can use the chart in FIG. 8-7 to find the correct value. In this case, follow the vertical line for the crossover frequency up to the horizontal line for impedance, then move diagonally to the upper right for the right value of inductance.

If you want to design a crossover network of higher order, go to TABLE 8-1 for the values. All you have to do is choose the type of network you want and use the multipliers in the chart to get the correct component values. If, for example, you decide to build an L-R second-order network for a set of 8-ohm drivers with a crossover frequency of 2000 Hz, first find the value of N from the chart in FIG. 8-7. The value of N is the same as that of a component for a first-order crossover. In this case, N would be 10 $\mu$F for the capacitor and 0.64 mH for the inductor. Use TABLE 8-1 to then find multipliers of 0.50 for both capacitors and 2.00 for both inductors. Thus, you would build a network using 5 $\mu$F capacitors and 1.28 mH coils.

## Series Networks

Series networks are not in fashion. The prejudice against them is such that few rally to their defense. One courageous advocate of series networks, Daniel Patrick Coyle, thinks they provide better transient clarity. He prefers them particularly for domes. I. M. Fried also favors series networks for his speaker systems. Much of the criticism of series networks is based on theory. Unfortunately, there has been little experimentation with controlled listening tests to decide this issue.

The first-order series network, shown in FIG. 8-1C, is in some ways a good choice for anyone who has no test equipment. A minor mistake in the value of a coil or a capacitor will have little effect on frequency response because it acts in a complementary way. That is, it shifts the frequency of the crossover point up or down to both woofer and tweeter instead of affecting one driver alone. The circuit is self-balancing.

It is possible to obtain different damping factors and crossover slopes with the first-order series circuit by adjusting L and C to values other than N. You can increase C and decrease L to make for a slightly sharper cutoff at the crossover frequency with some unevenness in response. Or L can be made greater while C is reduced for greater

**Table 8-2   Multipliers for First-Order
Series Networks**

| Circuit | L | C |
|---------|------|------|
| A | N | N |
| B | 1.23 | 0.83 |
| C | 2.00 | 0.50 |
| D | 0.71 | 1.41 |
| E | 0.50 | 2.00 |

damping. TABLE 8-2 shows some of the multiples for the N values that have been suggested by various theorists and experimenters.

The best choice among the possible combinations shown in TABLE 8-2 could depend on the kind of amplifier used. Amplifiers that are tolerant of either inductive or capacitive loads will work well with any one of the circuits. Some amplifiers give better performance with an inductive load than a capacitive load, while others have the reverse characteristic. This is one reason why some speaker systems and some amplifiers turn out to be highly compatible, while another combination is less pleasing.

For problem amplifiers, the best choice is circuit A, the ordinary first-order crossover using values from the chart or formulas. It has a constant input impedance that makes the load resistive rather than reactive. For maximum flat response from each speaker, choose circuit B. The other circuits, C, D, and E, have been recommended at various times because they can produce sharper roll-off at the crossover point than the normal circuit (A). Note that B and C will produce a more inductive load, while D and E will be more capacitive. The order in which the circuits are arranged, from A to E, is based on probable amplifier compatibility. Circuit A should be least amplifier sensitive; E is *probably* most likely to affect amplifier performance.

One requirement for the series network is that the drivers have equal impedance, at least at the crossover frequency. A typical woofer will have rising impedance that exceeds that of the typical tweeter at a typical crossover frequency. One way to compensate for this difference is to add resistance to the tweeter circuit. Most tweeters are more efficient than woofers and must be padded down. If you are lucky, the amount of resistance needed to equalize the impedances just might be the amount necessary to balance the drivers' outputs. Chapter 9 deals with other methods of controlling impedance variations.

There is one precaution to observe in using a first-order series crossover network. If you look at the circuit diagram in FIG. 8-1C, note that the tweeter is fed through the large capacitor, C, and shunted by the inductance, L. The inductor should offer a low impedance to low frequencies so that they flow through it and not through the tweeter. But if one uses an inductor with a higher-than-normal dc resistance, the tweeter might receive more of the bass signal than is desirable. For that reason, it might be wise to choose a coil with unusually low dc resistance for this circuit. Or, if you wind your own, use wire of larger diameter.

The circuits listed in TABLE 8-2 don't cover all the possible combinations. If you want to explore others, use these formulas:

$$R_W = R_T = R$$

$$L = R/(a2\pi f)$$

$$C = a/(R2\pi f)$$

where $R_W$ is the impedance of the woofer, $R_T$ is the impedance of the tweeter, and a is any value you want to use, within reason. Note that as you decrease a, you increase the value of L and decrease the value of C. As you change the value of a, the values of L and C must shift in a complementary way to maintain the desired crossover frequency.

## THREE-WAY CROSSOVER NETWORKS

Figure 8-6 shows two kinds of first-order three-way networks: parallel (FIG. 8-6A) and series (FIG. 8-6B). A three-way network adds another crossover point, which can cause problems, but it allows a bit more flexibility in choosing drivers. It also permits you to make finer adjustments of frequency response by using two L-pads, one for the midrange driver and one for the tweeter.

The first step in designing a three-way crossover network is to set the crossover frequencies. Your final choice depends to some extent on the kind of midrange speaker you choose. Some midrange drivers come with recommendations from the manufacturers on useful range. If you have no such specifications, you can usually estimate a proper low crossover point by considering the type of midrange unit you are using. When you use a small driver that was designed for use as a woofer, the crossover frequency can be rather low, at 300 or 400 Hz, for example. For cone-type midrange speakers, you can usually choose a 500- to 800-Hz crossover point. And with midrange domes, a crossover point of 700 to 1000 Hz is probably satisfactory.

The next step is to set the upper crossover frequency. Typically, the midrange driver covers about three octaves before crossing over to a tweeter. For example, the two crossover frequencies might be at 500 and 4000 Hz. This leaves somewhat more than two octaves to be covered by the tweeter. In some systems, the midrange driver's band is wider than three octaves.

In the case of the first-order parallel network as shown in FIG. 8-6A, the values of C1 and L2 can be obtained directly from the chart in FIG. 8-7 or from formulas. For 8-ohm speakers and crossover frequencies of 500 and 4000 Hz, those values turn out to be 5 $\mu$F and 2.5 mH. Note that you locate the value of inductance for 500 Hz because it is in the woofer circuit and the capacitance for 4000 Hz because it is in the tweeter circuit.

For the values of L1 and C2 in the bandpass circuit, you must make a slight calculation. Inductor L1 functions at the upper end of the midrange scale to limit highs to the driver, while C2 operates at the lower end. But because the components interact, some adjustment should be made to the standard values. The chart shows the value for an 8-ohm speaker at 4000 Hz to be 0.32 mH. In order to offset the interaction mentioned above, you should add about 13 percent to that value. This gives a final figure of 0.36 mH. For the capacitor with an 8-ohm impedance at 500 Hz, the chart lists 40 $\mu$F. Instead of using a 40 $\mu$F capacitor, subtract about 12 percent from that value, making it 35.2 $\mu$F. Note the rule: for a three-octave bandpass filter, add 13 percent to the inductor, and subtract 12 percent from the capacitor. In effect, you calculate the values for a slightly narrower band of frequencies than the desired band.

As you might have noticed, the bandpass in the example above was chosen so that the upper crossover frequency was eight times that of the lower one. In some cases, you might want to choose a different ratio of upper frequency to lower frequency, for example, 10 to 1. For a ratio of 10, you would add 11 percent to the value of the inductor and subtract 10 percent from the value of the capacitor.

If you wish to use another ratio, you can use the reciprocal of the ratio to approximate the fraction of the value of each component that must be added or subtracted. Note that the reciprocal of 10 is 0.1, or 10 percent.

To find the correct values of the components in the first-order, series, three-way network, as diagrammed in FIG. 8-6B, there are no extra calculations to make. The lower frequency crossover point is set by the two components L1 and C1, so these values can be obtained from the chart. For an 8-ohm system crossing over at 500 and

4000 Hz, these turn out to be 2.5 mH and 40 $\mu$F. The other components, L2 and C2, set the upper point. Those values are 0.32 mH and 5 $\mu$F.

## CROSSOVER NETWORKS FOR PIEZOELECTRIC TWEETERS

Piezoelectric tweeters have such a high impedance at low frequencies that they are effectively removed from the amplifier circuit. If the tweeter output level matches that of the other drivers, it can be wired directly to the amplifier output line with no crossover components to limit damping on the tweeter. These tweeters are usually more efficient than other drivers, requiring some kind of attenuation to balance them.

One way to use piezoelectric tweeters without a control is to install them at the back of the speaker system, firing at the rear wall. This disperses the upper highs and attenuates them so that a good balance can often be obtained.

Figure 8-8 shows how to add various series and shunt elements to control both the output level and the frequency response of piezoelectric tweeters. The circuit of FIG. 8-8C is particularly useful. These tweeters sometimes have a slight hump in response at the bottom end of their frequency band, and the series capacitor/shunt resistor circuit rounds off the response curve very nicely.

Some amplifiers react unfavorably to a capacitive load such as the piezo tweeters. For those amplifiers a series resistor, as shown in FIG. 8-8A, can stabilize the amplifier.

As you can see, the use of a piezoelectric tweeter requires very little design work. There is at least one kind of circuit, shown in FIG. 8-8D, in which you can adjust the value of C to raise or lower the crossover point. When an 8-ohm L-pad is added to the circuit, with a separate shunt resistor on the tweeter, you can calculate the proper value for C by assuming an 8-ohm impedance for the tweeter. Note that this circuit provides a continuously variable control for the tweeter, making it possible to match it to any other driver of lower efficiency.

## CROSSOVER NETWORK COMPONENTS

It is not always wise to substitute crossover network components of a different type than those specified in a published design. It is possible that the change will alter the efficiency or frequency balance of a system. On the other hand, if you like to experiment and are aware of

Circuit for high roll-off

| R | Effect at 20 kHz |
|------|------------------|
| 50Ω  | −2.5 dB |
| 100Ω | −5.0 dB |
| 150Ω | −7.5 dB |

**A**

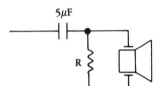

Low-end roll-off (3 dB/octave)

| R | Effect at 5 kHz |
|------|-----------------|
| 10Ω  | −2.50 dB |
| 8Ω   | −3.50 dB |
| 6Ω   | −4.25 dB |

**C**

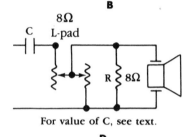

Attenuation without high roll-off

| C | Level |
|---------|-------|
| 1.0μf | −1 dB |
| 0.5μf | −2 dB |
| 0.2μf | −4 dB |
| 0.1μf | −7 dB |

**B**

For value of C, see text.

**D**

**8-8** Circuits for piezo-electric tweeters. Data for circuits A, B, and C from Piezo Ceramic Products Group, Motorola, Inc.

the problem, you can compensate for any changes in performance that might occur.

There are two general kinds of capacitors used in crossover networks, nonpolarized (NP) electrolytic and solid dielectric capacitors. Solid dielectric, such as Mylar or polypropylene capacitors, are more stable and behave more like a pure capacitance at high frequencies than NP electrolytics. All capacitors contain elements of resistance and inductance as well as capacitance. These "impurities" make the capacitor act like a resonant circuit at some high frequency. Below that critical frequency, which varies according to the type of capacitor, the capacitor acts more or less like a true capacitance; above that frequency it acts like an inductor. At the resonance frequency, the capacitor is resistive. In general, you could say that the lower the frequency, the better a capacitor "acts."

In view of the characteristics mentioned above, it might seem that no NP electrolytics should be used for crossover duty. That is not true. Good NP electrolytics are perfectly satisfactory, especially for low-frequency duty in woofer circuits. They are also usually good enough for tweeter circuits, particularly if they are bypassed by putting a 0.1 $\mu$F Mylar or polypropylene capacitor in parallel with the electrolytic. Bypassing the NP electrolytic improves high-frequency transient performance.

Be aware that there is a difference in the efficiency balance of a system using an NP electrolytic in the tweeter circuit and a system that is identical except for a solid dielectric type of capacitor feeding the tweeter. The system with the solid dielectric capacitor will sound brighter because of greater tweeter output. If there is a tweeter control in the crossover circuit, this kind of change poses no problems. You can reduce the drive to the tweeter by adjusting the L-pad.

In some cases, random selections of various kinds of capacitors of the same specifications show differences in value. I have found NP electrolytics to run slightly lower in value than typical Mylar capacitors, for example. Some others agree on this difference, and some say that the opposite condition is true. If you have facilities for measuring crossover component values by one of the tests described in Chapter 10, that is worth doing. Some mail-order houses supply capacitors that are matched to within 1 percent for a small fee.

Another way capacitors differ is in voltage rating. For almost any conceivable situation if you choose capacitors that have a voltage rating of 50 volts or better, you won't have problems. In an 8-ohm system, such a capacitor is safe up to 150 watts.

Two kinds of inductors are used in speaker system crossover networks: air-core coils and iron-core coils. An air-core coil can have a solid core if the core is made of nonmagnetic material, such as wood or plastic. When an iron core is used, a coil can have the same inductance value with less copper wire. It is desirable to keep the dc resistance of a coil as low as possible, but it is questionable whether using an iron core is the way to do that. Cheap iron-core coils are likely to produce distortion. Better iron-core coils are sometimes specified for woofer circuits where the dc resistance of an air-core coil might cause a problem. As a rule of thumb, it is desirable to use a coil that has a dc resistance of no more than 5 percent of the impedance of the driver. Low-resistance air-core coils must be wound with copper wire of larger diameter than coils with higher resistance, so they are more expensive. In almost all cases, it pays to get the best air-core coil you can afford.

L-pads that are used in the midrange and tweeter circuits should be chosen to have a power rating to match their job. Ordinary music contains far less power in the high-frequency part of the spectrum than in the low or middle parts, so you can use a lower-rated L-pad for the tweeter than for the midrange speaker. Typically, a 15- or 25-watt L-pad is fine for a tweeter circuit, but the midrange L-pad should be rated at 50 watts or better.

Fixed resistors can be used in place of L-pads but lack the flexibility of changing the balance of a system when it is moved from one room to another. They are used in some of the projects in this book, such as the compact system with a tile back, where adding an L-pad would have required an extra hole in the tile back.

## CROSSOVER CONSTRUCTION AND INSTALLATION

The parts of your crossover can be mounted on a small piece of hardboard or pegboard. Try to separate coils by at least 2 inches, if possible. If two coils must be placed near each other, mount them so that they are at right angles to each other to reduce the chances of interaction. Don't let resistors touch other parts. After wiring and soldering the connections, glue every part to the board with silicone rubber or other adhesive.

Most crossover networks are installed in the speaker enclosure. If you do that, make sure the network is located as far as possible from the drivers. It is possible for magnetic fields of the drivers and the crossover components to interact and cause distortion. If you install the board with screws, use brass screws.

## WHICH KIND OF CROSSOVER NETWORK IS BEST?

There is no one kind of crossover network that is best for all situations. A small tweeter used in a speaker system that will probably be driven to high sound intensity levels might, depending on the crossover frequency, require a higher order network than a similar tweeter in another system. With that qualification, it is generally good policy to say "the simpler, the better."

If you buy components off the shelf and use them as if their values of inductance and capacitance are exactly as marked, there will undoubtedly be some variation from the theoretical ideal. The greater the number of parts in your crossover network, the greater the possibility that some of the variations will be additive, multiplying the effect of the errors. Add to that the possibility of "ringing," or oscillation, which is greater for networks of higher order. It seems that there "ought to be a law," such as:

$$CPT = 1/N^2$$

where CPT is the chance of a network performing according to theory and N is the number of components in the network.

Apparently, given the kinds of drivers available today, we must have crossover networks. When you consider the problems of inter-action, frequency response variations, and phase differences produced by using separate drivers, you might wonder why more research hasn't been done on producing full-range drivers. Such drivers are appropriate only for low-cost speaker systems. Recently I had an occasion to compare the performance of a 6-×-9-inch extended range car speaker purchased a few years ago with a newer, cheap coaxial 6-×-9-inch car speaker. The older speaker with a single cone gave better performance in every measurable way except upper high frequency response. In fact, the cheap coax was hard to listen to. Why make such a speaker? Apparently because the average buyer is attracted by multiple drivers regardless of quality.

Whatever crossover you employ, you can probably improve the performance of your system by tweaking it. This is an argument for installing the crossover outside the enclosure so you can experiment with alternate values of components. If, after careful listening, you prefer a change from a "book" crossover, trust your ears.

## PROJECT 9: Three-way Speakers

Here is a three-way system that consists of an American-made 10-inch woofer, a 5-inch cone midrange, and a 1-inch textile dome tweeter (TABLE 8-3). The midrange speaker and the tweeter are made in Europe by Philips. They are highly efficient models that demand L-pads to match their output to that of the woofer.

The enclosure is a standard box design, so standard that it could be used with any number of 10-inch woofers. Ed Grundy, a skilled cabinetmaker, built the cabinet shown in the photograph (FIG. 8-9). He added a base and a panel with a shaped edge to cover the lower compartment. As designed, the cabinet is a speaker box set on a built-in stand with a lower compartment for the crossover network (FIG. 8-10). You can eliminate the bottom structure, but for best performance you would need a separate stand. In that case, you could install the crossover network in the back of the speaker enclosure.

The material used for the cabinet here was ¾-inch high-density particle board with oak veneer and a hard lacquer finish. Because you

**Table 8-3 Parts List for Project 9: Three-Way Speakers**

| *Pieces* | *Dimensions* | *Function* |
|---|---|---|
| 2 | ¾ × 13⅝ × 33″ plywood | Sides |
| 2 | ¾ × 13⅚ × 12¼″ plywood | Top & bottom |
| 1 | ¾ × 10¾ × 22½″ particle board | Back |
| 1 | ¾ × 10¾ × 31½″ particle board | Speaker board |
| 1 | ¼ × 10¾ × 31½″ plywood | Faceplate |
| 1 | ¾ × 10¾ × 12⅛″ | Partition |
| 1 | ¼ × 10¾ × 8¼″ plywood | Control board |
| 16 ft. | ¾ × ¾″ hardwood | Cleats & braces |
| 4 ft. | ¾ × 2¼″ hardwood | Base (optional) |
| | **Components** | |
| 1 | 10″ woofer | Eminence EM-40 |
| 1 | 4½″ midrange driver | Philips AD50600SQ8 |
| 1 | 1″ soft-dome tweeter | Philips AD11600T8 |
| 1 | 4 uF capacitor, Mylar, 100 V | C1 |
| 1 | 24 uF capacitor, NP electrolytic, 100 V | C2 |
| 1 | 4.7 uF capacitor, NP electrolytic, 100 V | C3 |
| 1 | 22 uF capacitor, NP electrolytic, 100 V | C4 |
| 1 | 6 uF capacitor, NP electrolytic, 100 V | C5 |
| 1 | 8 Ω resistor, 10 W | R1 |
| 1 | 7.5 Ω resistor, 15 W | R2 |
| 1 | 0.3 mH inductor, air core | L1 |
| 1 | 1.8 mH inductor, air core | L2 |
| 1 | 1 mH inductor, air core | L3 |
| 1 | 8 Ω L-pad, 15 W | Tweeter control |
| 1 | 8 Ω L-pad, 50 W | Midrange control |
| 1 | Speaker terminal plate | |

might not want to apply veneer, the parts list in TABLE 8-3 shows ¾-inch plywood for the enclosure. Note that you have to cut away the speaker board to within less than an inch of each edge to install the woofer. To prevent breakage, glue a sheet of ¼-inch plywood to the panel. Ed Grundy installed the drivers by routing the board to the proper depth and shape for each one. If you have no facility for routing, you can cut out the plywood panel to fit around the outside of each driver frame.

To install the crossover network and controls in the bottom compartment, it is necessary to route the driver cables into the speaker compartment through the partition. One way to obtain an airtight installation is to drill holes just large enough to pass the cable

**8-9** Project 9: Three-Way Speakers.

through, then fill the space around the cable with silicone rubber sealer. Another possibility is to install terminals in the partition, but this makes just one more place for corrosion to interfere with signal transmission.

The box should be well braced. If possible, line the walls with asphalt roofing material as described in Chapter 4. Cover the interior walls with damping material and the back with a double layer. For the cabinet shown in the photograph, a 2-inch layer of polyester batting, held in place with nylon netting, was placed inside the back panel.

The crossover network used here is a standard book design with crossover points at 700 and 5000 Hz, plus some additions to control impedance and response behavior of the woofer and midrange driver (FIG. 8-11). The only real response problem was a peak in the woofer's output above its operating range, which explains the special filter, consisting of L3 and C5, in series with it. (Such filters are

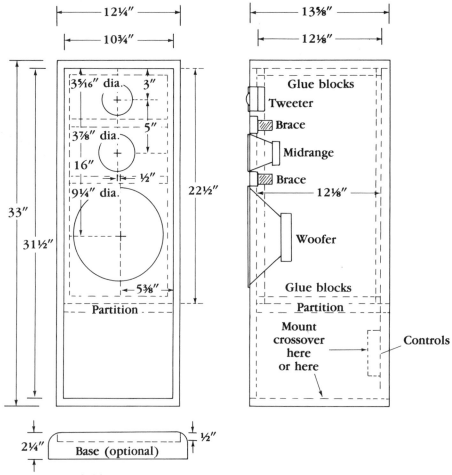

**8-10**   Construction plans for Project 9 enclosures.

described in Chapter 9.) If you substitute another woofer, that filter can probably be removed.

Once the problem with the woofer was solved, this became a very satisfactory system. It reproduces piano music with authority, the larger woofer rendering the bass tones of a grand piano quite realistically. If you build a couple of these, don't forget to use adequate bracing.

## PROJECT 10: Double-chamber Tower Speakers

This project consists of an 8-inch woofer and a 1-inch tweeter in a double-chamber reflex tower (FIG. 8-12 and TABLE 8-4). The cabinet

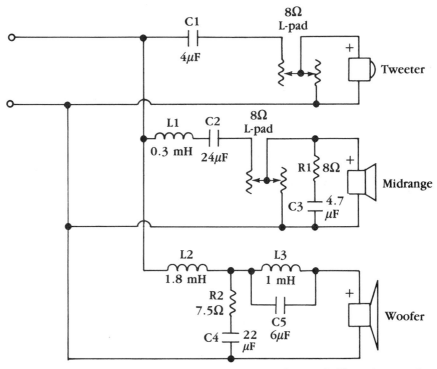

**8-11** Diagram of crossover network used in Project 9. If another woofer is substituted, omit L3 and C5.

dimensions are 10 inches wide by 40 inches tall by 12¼ inches deep. You can shorten the towers somewhat if desired because the bottom compartment houses only the crossover network and has more space in it than is absolutely necessary.

You might notice that the cabinets in the photographs have shaped edges on the top and bottom birch plywood panels. A simple square edge will do there, but you might want to cover the edges with thin strips of wood veneer.

The dimensions shown in FIG. 8-13 are the measurements in the enclosure before adding a layer of asphalt roofing material and a layer of thin, foam-backed carpet to the interior sides, top, bottom, and back by the method described in Chapter 4. Note that the speaker board is a combination panel made up of a sheet of ¾-inch particle board and ¼-inch plywood. To make it, cut out the holes for the drivers in the particle board and set the woofer in position. Use care in handling the board; with so little material left at each side of the woofer hole, it is easily broken. Rotate the woofer until the four

Hands-On Electronics

**8-12**   Project 10: Double-Chamber Tower Speakers.

mounting holes are at the 2, 4, 8, and 10 o'clock positions. Mark these locations and drill ¼-inch holes to accept T-nuts at the rear. Do not cut the port holes at this time.

Next, cut the larger holes in the plywood faceplate. When you cut the woofer hole, the saw will cut the panel in two at the indicated breakpoints (FIG. 8-14). After making the tweeter hole cut-out, round off, or bevel, the edges of the faceplate around the tweeter hole (FIG. 8-15). Glue the faceplate to the speaker board. If you have clamps, use them to hold the panels while the glue sets. If you have no clamps, small nails or screws will do the job.

Next, cut the port holes. If you have no tools for cutting a perfect circle, you can use a port with a square cross-section. For the latter,

**Table 8-4  Parts List for Project 10: Double-Chamber Tower Speakers**

| Pieces | Dimensions | Function |
|---|---|---|
| 2 | ¾ × 12 × 38½″ plywood | Sides |
| 1 | ¾ × 12¼ × 10″ plywood | Bottom |
| 1 | ¾ × 8½ × 32½″ particle board | Back |
| 1 | ¾ × 8½ × 38½″ particle board | Speaker board |
| 1 | ¼ × 8½ × 38½″ plywood | Faceplate |
| 1 | ¾ × 8½ × 10″ particle board | Partition |
| 30 ft. | ¾ × ¾ hardwood | Cleats & braces |
| 10 ft. | ¼ × 1⅛″ lattice wood | Grille frame |
| 3 | 2 (I.D.) × 5½″ PVC pipe | Port tubes |
| | **Components** | |
| 1 | 8″ woofer | Peerless 1556 |
| 1 | 1″ soft dome tweeter | Polydax HD100D25 |
| 1 | 5.5 uF capacitor, Mylar, 100 V | C1 |
| 1 | 10.5 uF capacitor, Mylar, 100 V | C2 |
| 1 | 12 uF capacitor, NP electrolytic, 100 V | C3 |
| 1 | 0.35 mH inductor, air core | L1 |
| 1 | 0.5 mH inductor, air core | L2 |
| 1 | 8 Ω resistor, 15 W | R1 |
| 1 | 8 Ω L-pad, 50 W | Tweeter control |
| 1 | Speaker terminal plate | |

*Hands-On Electronics*

make the opening 1¾×1¾ inches and to the same length as that shown in the drawing (5½ inches).

From here on, the assembly follows standard procedure. Use plenty of bracing. A ¾-×-¾-inch brace should go on the back of the speaker panel between the woofer and tweeter holes. The back should have at least one edge-on brace placed vertically, just off center. Even better, cross bracing, with the back tied to the other parts, would be good.

Note that the parts list (TABLE 8-4) specifies Mylar capacitors in the tweeter branch of the crossover network (FIG. 8-16). To get a 5.5 $\mu$F capacitor, you can wire a 2.2 $\mu$F capacitor in parallel with a 3.3 $\mu$F capacitor (both easy values to obtain). For the 10.5 $\mu$F value, wire 6.8, 2.2, and 1.5 $\mu$F capacitors in parallel. If you have facilities for measuring capacitance, you might find other combinations that will

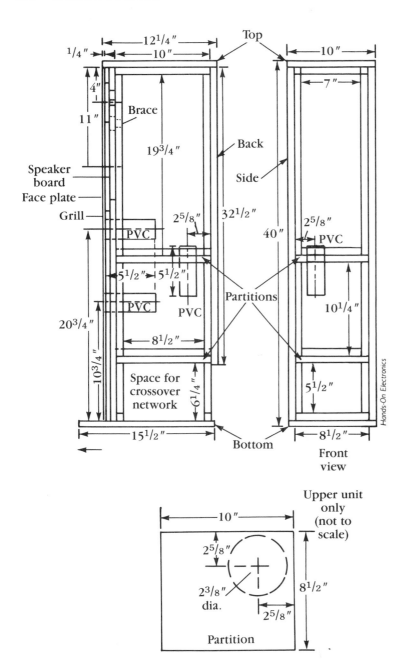

**8-13** Construction plans for Project 10 enclosures.

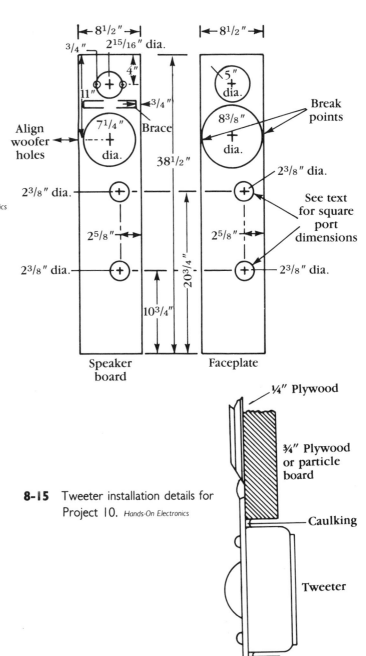

**8-14** Details of speaker board (left) and faceplate (right) for Project 10. *Hands-On Electronics*

**8-15** Tweeter installation details for Project 10. *Hands-On Electronics*

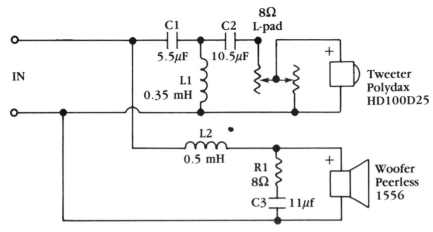

**8-16** Diagram of crossover network for Project 10. *Hands-On Electronics*

**8-17** Internal construction of Project 10 enclosure. Note use of a clamped board to protect cutout speaker board during work on project.

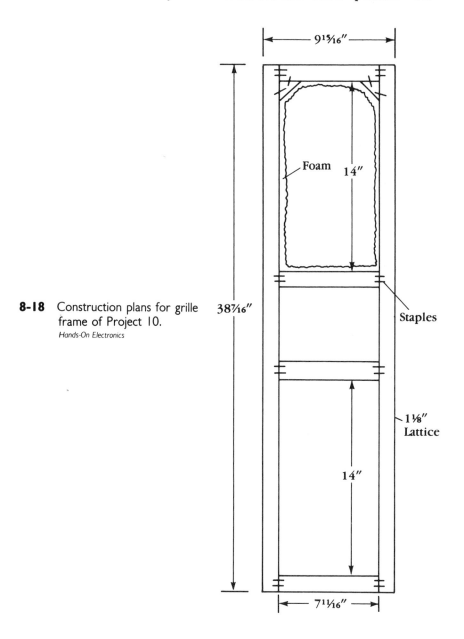

**8-18** Construction plans for grille frame of Project 10.
*Hands-On Electronics*

work. You can substitute a 12 $\mu$F NP electrolytic for the 11 $\mu$F value of C3 in the woofer impedance equalizer specified by Peerless.

Mount the crossover network on a board and install it in the lower compartment. Make sure you have an airtight passage through the partition when you run wiring to the drivers.

Line the interior of the large compartment, except for the speaker board, with a 1-inch layer of fiberglass batting (FIG. 8-17). Use extra on the back behind the woofer, but keep the entrance to the ports free.

Install the tweeter with silicone rubber sealer and small screws. Glue a ring of foam or felt around it in the depression that surrounds the frame. It's a good idea to install the woofer with a gasket of foam weathering tape behind it so that the driver can be easily removed if necessary. For example, if you hear any coloration, you may want to rearrange the fiberglass damping material or add some loose polyester batting behind the woofer. If you are satisfied that everything is working well, remove the woofers and glue them down with silicone rubber sealer. You can even fill the bolt holes with the sealer and have a semi-flexible mounting for the woofers. Allow sufficient time, up to 24 hours, for the silicone rubber to set before using the speakers.

Make a frame out of ¼-×-1⅛-inch lattice material to hold the grille cloth (FIG. 8-18). You can install the grille with velcro strips or even small brads.

# Chapter 9

# How to Tame Peaks

$W$hen you start to design a speaker system, you make an assumption that the drivers have a flat frequency response. Some are flatter than others, but none are perfect. Another assumption that consumers sometimes make is that the figure the manufacturer lists for speaker impedance is a constant. That, of course, isn't true.

## HOW SPEAKER IMPEDANCE VARIES

To see how a speaker's impedance can vary with frequency, see the solid-line impedance curve in FIG. 9-1. This curve is typical of many woofers. The low-frequency peak occurs at the fundamental resonance where the cone motion is at a maximum, producing a back EMF. Above the resonance frequency, the impedance drops to, or below, its rated value. In the case shown in FIG. 9-1, the minimum impedance occurs at about 100 Hz. For smaller drivers, the minimum occurs at 200 to 400 Hz. Above that trough, the impedance begins to rise because of the inductive reactance of the voice coil.

An electric current varies inversely with the impedance of the load. Looking at FIG. 9-1 you might expect the driver to have a very limited response range with an electrically induced cutoff in the bass as well as at high frequencies. The rising impedance works that way at the high frequencies, but not in the bass. As the frequency is lowered, the driver coil passes less current, but this loss of drive is matched by an increase in efficiency. Efficiency peaks at resonance.

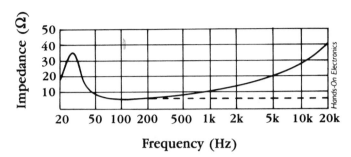

*Hands-On Electronics*

**9-1**   Impedance curves for a 10-inch woofer before (solid line) and after an impedance equalizer was installed (dashed line).

Figure 9-1 shows one reason why large woofers, particularly those with long voice coils and greater voice coil inductance, cannot be used as full-range speakers. The inductance of the voice coil, $L_E$, is too great. In fact, you can make an estimate of the upper frequency limit of a woofer by considering the effects of voice coil inductance by this formula:

$$f_H = R_E/(2\pi L_E)$$

where $f_H$ is the expected high frequency limit, $R_E$ is the dc resistance of the voice coil in ohms, and $L_E$ is the inductance of the voice coil in henries. For example, if a woofer has an $L_E$ of 1 mH and an $R_E$ of 6 ohms, its high cutoff would be:

$$f_H = 6/[2\pi(0.001)]$$
$$= 955 \text{ Hz}$$

This is not an absolute figure; the real situation is more complicated. It can be taken as a rough guide to the kind of high frequency response one can expect. For a rule of thumb that works fairly well, double the frequency obtained by the formula above to estimate the upper frequency limit of a woofer.

Impedance variations cause a problem for the crossover network. The solution is an impedance equalizer.

## HOW TO DESIGN IMPEDANCE EQUALIZERS

You can equalize the impedance rise caused by voice coil inductance by connecting a simple network across the voice coil (FIG. 9-2). If

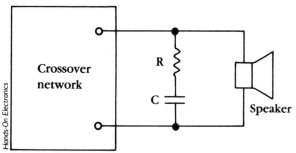

**9-2**   This impedance equalizer flattens the impedance curve and, if designed properly, can enhance the crossover's performance.

you know the dc resistance of the voice coil, $R_E$, and inductance, $L_E$, you can calculate the value of the capacitance by:

$$C = L_E/R_E{}^2$$

and the resistance by:

$$R = 1.25\ R_E$$

Here is an example: the specification sheet for a small woofer shows $R_E$ to be 6.7 ohms and $L_E$ to be 0.5 mH, or 0.0005 H. Therefore:

$$C = 0.0005/44.89$$

$$= 11\ \mu F$$

and, for R:

$$R = 1.25 \times 6.7\ \text{ohms}$$

$$= 8.4\ \text{ohms}$$

The formula above for C is a textbook formula that gives a value that is greater than the one recommended by some manufacturers. One would expect the company that makes a driver to recommend the optimum value of components to be used with that driver. With that in mind, here is a rule-of-thumb test that yields lower values for C. It can be used for any woofer you might have that has no specification for $L_E$. First, measure the dc resistance of the voice coil with an ohmmeter. Record the value as $R_E$. Then run an impedance curve by Test 2 as described in Chapter 10. Find the point on the curve where the impedance equals twice the value of R. Record the frequency of that point as f. Then find the value of C by:

$$C = 0.16/(R_E \times f)$$

where f is the frequency noted above.

To see the effect of an equalizer on the impedance curve of a woofer, look at FIG. 9-1 again. The solid line is the curve of the woofer alone, and the dashed line is with an impedance equalizer installed.

You can use impedance equalizers with midrange drivers and tweeters too. Figure 9-3 shows the impedance curve for a titanium-dome midrange driver, the MCD 51M. The curve for this German-made driver shows the original impedance (solid line) and after two circuits were added (dashed line). The two networks were an *impedance equalizer* (which leveled the high frequency curve) and a *resonant-peak filter*.

Here are some guides: the value of C for woofers will usually fall in the 10- to 50-uF range; for midrange domes, 2 to 5 uF; and for midrange cone drivers, 3 to 8 uF. You can usually get good performance from tweeters without equalizers, but their sound is sometimes smoother if one is used. For tweeters, C usually varies from 1 to 2 uF. If you measure $R_E$ for small tweeters, use a digital ohmmeter to limit the current applied to them. If you have no digital ohmmeter, use an 8-ohm resistor in the impedance equalizer for an 8-ohm tweeter.

If you use an impedance equalizer with your woofer, consider its effect in designing a crossover network. Typically, you can use $R_E$ as the impedance of the woofer in calculating crossover component values. For a midrange driver or a tweeter, the L-pad, if you use one, helps determine the proper impedance value to use. With an 8-ohm L-pad on a typical midrange driver or tweeter, the final impedance will be close to 8 ohms.

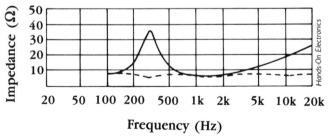

**9-3**  Impedance curves of a midrange dome before (solid line) and after an impedance equalizer and a resonant-peak filter were installed (dashed line).

## RESONANT-PEAK FILTERS

Tweeters and midrange drivers often present a problem at their fundamental resonance. This occurs at the lower limit of the frequency response range. Unlike the woofer's fundamental resonance frequency, that of a higher frequency driver occurs in its stopband, where it can upset the crossover network's performance. Whether this causes a problem depends on how far the crossover point is removed from the resonance and how well that resonance is damped by driver design. Drivers with Ferrofluid in the magnetic gap usually pose no problems. To reduce the possibility of a peak in response, you can add a series-tuned circuit that resonates at the same frequency as the impedance peak. Figure 9-4 shows the schematic diagram for such a filter. To design a resonant-peak filter, find the frequency of the peak at resonance for your midrange driver or tweeter, then use this formula:

$$C = 1/(50f)$$

This formula assumes that you are using an 8-ohm tweeter. If your tweeter has an impedance of 4 ohms, substitute 25f in the denominator.

To find the inductance that will resonate with C at the peak frequency, use this formula:

$$L = 0.025/(f^2C)$$

For R, choose a value equal to the rated impedance of the driver or slightly higher.

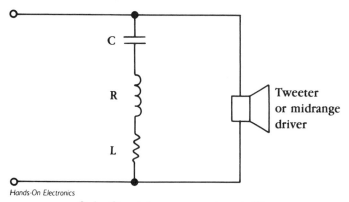

*Hands-On Electronics*

**9-4**  Circuit for a resonant-peak filter.

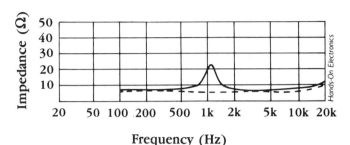

**Frequency (Hz)**

**9-5**  Impedance curves for a small dome tweeter before (solid line) and after a resonant-peak filter was installed (dashed line).

Figure 9-5 shows the impedance of one tweeter before (solid line) and after a resonant-peak filter was installed (dashed line). The tweeter, a German-made MCM 25S had noticeable output at its resonance point of 1.1 KHz, even though it was connected to a third-order network that crossed over at 2.2 kHz. The resonant-peak filter used in this case had an 18 uF capacitor in series with a 1.14 mH coil and an 8-ohm resistor.

The formulas shown here work well without requiring the detailed testing procedures of more specific formulas. Remember that such formulas are only a guide to reasonable values of the components of such filters. You can adjust the values by experiment to obtain a flat impedance curve. If you don't have the exact value of inductor but do have a similar one, use the formula to calculate a different value of C that will resonate at the desired frequency. It is often useful to choose L first because odd values of coils might not be available. For example, suppose you have a coil with an inductance of 1.25 mH and you want to make a filter for 1.1 kHz. You would use the resonance formula:

$$C = 0.025/(f^2L)$$

In this case, you need a 16.5 uF capacitor.

Peak filters for midrange drivers can require rather large values for L. The filter for the MCD 51M (impedance curve is shown in FIG. 9-3), for example, requires an inductance of 4.15 mH.

The formulas listed here give an LC ratio of about 63. You can substitute lower values for L along with higher values for C and get good results. If possible, run an impedance curve on your corrected driver. If the impedance falls below 3 or 4 ohms at any point, increase the value of R to 10, 12, or even 15 ohms.

Remember that resonant-peak filters are not always needed. But in some cases, where resonance of a midrange driver or a tweeter is a problem, they can clean up a system.

## NOTCH FILTERS

A notch filter is a frequency-discriminating circuit consisting of an inductor, a capacitor, and, usually, a resistor. The components are wired in parallel with each other with the network in series with the load (FIG. 9-6). The inductive reactance must equal the capacitive reactance at the frequency of the peak you want to control. The resistor controls the sharpness and magnitude of the filter's response. The higher its resistance, the greater the cut in speaker response.

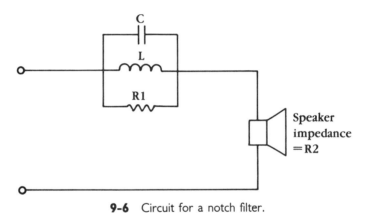

**9-6**  Circuit for a notch filter.

When you compare inductance and capacitance values in various notch filters, you are likely to be puzzled by the design choices. After all, the range is almost infinite as long as the capacitive reactance equals inductive reactance at the peak frequency. You could attempt to understand filter action better by going back to basic engineering textbooks, but you might find you have as many questions after reading them as you had before. The textbooks deal only with theoretical conditions where each component is perfect.

If you have an 8-ohm speaker that has a peak in its response range that should be reduced, you can use this formula for the value of capacitor in the filter:

$$C = 1/(33f)$$

where f is the frequency of the peak. If the impedance of your problem speaker is 4 ohms, substitute 16.5f in the denominator of the equation above. After you find the value C, you can find the right inductance to make the filter resonate at the desired frequency by:

$$L = 0.025/(f^2C)$$

Here is an example. An 8-ohm driver has a peak at 1000 Hz; what values of C and L would be appropriate to nullify the peak?

$$C = 1/(33f)$$
$$= 1/(33 \times 1000)$$
$$= 0.00003 \, F \text{ or } 30 \mu F$$

and

$$L = 0.025/(f^2C)$$
$$= 0.025/(1000^2 \times 0.00003)$$
$$= 0.00083 \, H \text{ or } 0.83 \, mH$$

Next, choose the right value of R to control the action of the filter at resonance. Before choosing R, you should check the Q of the peak. If the manufacturer of the driver furnishes a frequency response test, you can locate the points at each side of the response peak where the output is down by 3 dB, or do Test 7 in Chapter 10. Then you can find Q by:

$$Q = f_p/(f_H - f_L)$$

where $f_p$ is the frequency of the peak, $f_H$ is the frequency above and $f_L$ the frequency below the peak where the response is down 3 dB. You can then find R by:

$$R = Q/\sqrt{C/L}$$

Using the example above, if the peak has a Q of 5, then you could find R by:

$$R = 5/\sqrt{0.00003/0.00083}$$
$$= 26.3 \text{ ohms}$$

Note that the generalized design process above assumed an 8-ohm driver. Ralph Gonzalez of Delaware Acoustics suggests a more specific design that takes into account the impedance of the load. Figure 9-6 shows the speaker load designated by R2. Gonzalez notes that the notch filter's cut will go to zero if R2 is very great, but the cut will increase as R2 is reduced. To use the load impedance in calculating component values, Gonzalez suggests these formulas:

$$R1 = R2Q - R2$$

$$C = Q/(2\pi fR1)$$

$$L = R1/(2\pi fQ)$$

## How Component Quality Can Affect Notch Filters

Some textbooks state that the Q of a parallel resonant circuit is determined by the Q of the coil as well as by the parallel resistance. The capacitor, it is said, has little or no effect on Q. These statements might be more valid for high-frequency circuits than for the audio band. I checked their relevance to audio notch filters by running tests on some notch filters made up with a single inductor paired first with one kind of capacitor and then another. In one filter, when I used an NP electrolytic capacitor and a coil with no parallel resistor, I found that the maximum Q was 5.4 (TABLE 9-1 and FIG. 9-7). After

**Table 9-1  Theoretical Q, vs. Measured Q, of Notch Filters**

|  |  | Measured Q | |
|---|---|---|---|
|  | *Theoretical* | *NP Elect.* | *Mylar* |
| *R* | *Q* | *Cap.* | *Cap.* |
| 1.0 | 0.42 | 0.48 | 0.47 |
| 1.5 | 0.63 | 0.63 | 0.63 |
| 2.5 | 1.05 | 0.95 | 0.95 |
| 5.0 | 2.10 | 1.70 | 1.80 |
| 10.0 | 4.20 | 2.60 | 3.00 |
| 15 | 6.20 | 2.90 | 3.90 |
| 22 | 9.10 | 3.50 | 4.80 |
| 30 | 12.50 | 3.70 | 5.50 |
| 100 | 41.50 | 4.30 | 8.10 |
| 510 | 212.00 | 5.20 | 10.30 |
| ∞ | ∞ | 5.40 | 12.00 |

$C = 21.4\mu F$; $L = 0.13mH$; $f = 3kHz$. LC ratio = 6.

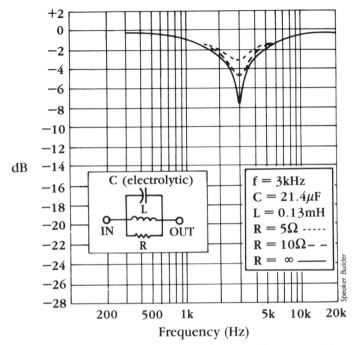

**9-7**   Frequency response of a notch filter made with an NP electrolytic capacitor.

replacing the NP capacitor with several Mylar capacitors, paralleled to make a capacitance of the same value, I found that the maximum Q was 12 (TABLE 9-1 and FIG. 9-8). These tests suggest that the real response of a filter may be quite different from the theoretical response, depending on the kind of components used in the filter. For the tests reported in FIG. 9-7 and FIG. 9-8, the coil in the filter was made with #20-gauge magnet wire and had a dc resistance of about 0.2 ohms. A coil made with wire of larger diameter would no doubt have produced a filter with a higher Q than the one shown in the graphs.

If you have test equipment, test the response of any notch filter you make. For a test procedure, refer to Test 16 in Chapter 10.

If you use a notch filter in the frequency range of a driver where the driver's impedance varies from the rated value, you might want to use an impedance equalizer on the driver. The impedance equalizer should be designed and installed so that the modified impedance value can be used in calculations for the notch filter.

Ralph Gonzalez suggests the use of a series resonant circuit between the crossover network and the notch filter (FIG. 9-9). This

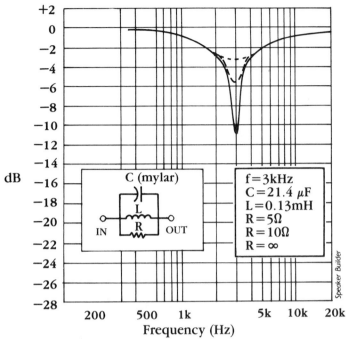

**9-8** Frequency response of the same notch filter reported in Fig. 9-7, but with a Mylar capacitor.

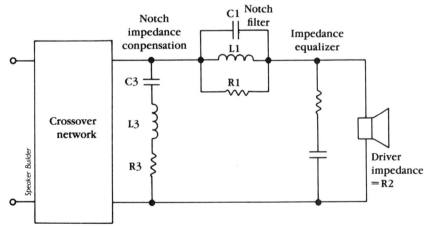

**9-9** Ralph Gonzalez of Delaware Acoustics suggests this notch impedance compensation network when the notch frequency occurs near a crossover frequency.

notch impedance compensation would avoid making the crossover see a high impedance at the notch frequency, which could upset the performance of the crossover network. If the notch frequency is close to the crossover frequency, the Gonzalez circuit is particularly worthwhile.

To choose the components for the notch impedance compensation network in FIG. 9-9, use these formulas:

$$C3 = L1/R2^2$$

$$L3 = R2^2C1$$

$$R3 = R2 [1 + (R2/R1)]$$

## SPECIAL CIRCUITS FOR TWEETERS

Some low- to medium-priced dome tweeters have a peak in the upper part of their audible range, making them sound "sizzly." One way to compensate for this characteristic is to use a notch filter, as described above. Another problem, common with many low-priced tweeters, is that there is too much output at the lower end of their response range combined with a falling response in the upper frequencies. The normal cure for this condition is shown in FIG. 9-10. For a typical cheap tweeter with falling response above 10 kHz, the value of the capacitor should be about 2 $\mu$F. The value of the resistor should be about equal to the tweeter's impedance. Use a larger capacitor to lower the corner frequency of the filter and a smaller one to raise it.

G. R. Koonce, in the March 1981 issue of *Speaker Builder* magazine, suggested the use of a simple series resistor to correct a tweeter's drooping high-frequency response (FIG. 9-11). As he noted, tweeters as well as woofers can have the problem of rising impedance at high frequencies because of voice coil inductance. By adding resistance to what is already a series resistive-inductive circuit, you can raise the turnover frequency. As Koonce noted, this not only

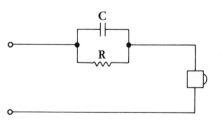

**9-10**   Circuit to boost the upper high frequencies of a tweeter.

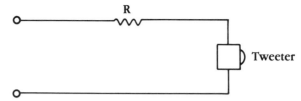

**9-11** Using a simple resistor in series with a tweeter, as suggested by G. R. Koonce, boosts upper highs, balances the tweeter to the woofer, and requires a reduced value of capacitor in the tweeter circuit.

saves the capacitor shown in FIG. 9-10, but it also allows you to use a capacitor of smaller value in the high-pass leg of the crossover network. With the use of a smaller capacitance, it is more practical to choose a Mylar capacitor for that role instead of an NP electrolytic. If you find that the simple resistance boosts the upper frequencies more than is desirable, you can use a combination of series and parallel resistors to get the kind of performance you want. Don't forget to use the total impedance reflected to the crossover network to calculate the value of the high-pass capacitor or that of any other crossover components in the tweeter leg of the circuit.

In all of these suggested modifications, you should make the final choice of parts by a listening test. But be aware that the listening test with and without any frequency-correcting filter should be made at the same loudness level. If this isn't done, your decision on the usefulness of the filter will be influenced as much by the change in loudness as by the frequency action of the filter.

## SPECIAL CIRCUITS FOR WOOFERS

Many small speaker systems sound thin because of a rise in response at some frequency related to the width of the enclosure. This effect is worth considering if your small speaker sounds too "forward" to suit you.

The theoretical frequency of the step response can be found by dividing the number 13,500 by the width of the enclosure in inches. Thus, a cabinet 10 inches wide would be expected to have a step response at 1.35 kHz.

Ralph Gonzalez of Delaware Acoustics suggests using the circuit shown in FIG. 9-12 as a corrective measure. To use it, choose the resistor, R, to equal the impedance of the woofer. You can find the value of L2 from the design chart or formulas in Chapter 8, but select it as if you are designing a first-order filter for your woofer at f/3,

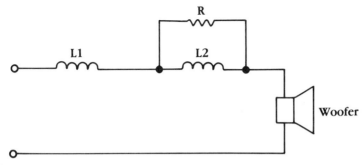

**9-12**   This circuit can be used to boost lower bass response. (See text.)

where f is the step frequency. This circuit will require about 6 dB extra cut in the tweeter drive.

Another way to defeat the effect of the step response is to use a woofer with a double voice coil (FIG. 9-13). One voice coil carries only the low bass. Choose the inductor for the bass-only voice coil, L2, as if for a crossover frequency of f/6. The value of L1 is the normal one for your woofer at the crossover frequency.

You can simulate a double voice coil woofer system by using two front-mounted woofers. The same rules apply here as for a double voice coil woofer, but you have the extra cone area for greater power handling.

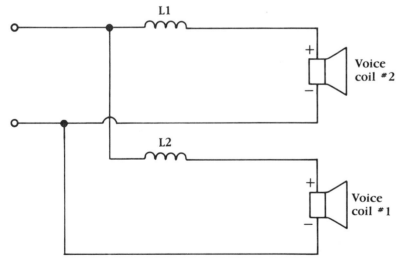

**9-13**   When a woofer with a double voice coil is used, one voice coil can be fed with a larger-than-normal inductance, boosting the lower bass. If voice coil #1 is the boost coil, L2 might have several times the value of L1.

**9-14**   Project 11: Double-Woofer Speakers.

## PROJECT 11: Double-Woofer Speakers

This project has two unusual features: the double woofers and an enclosure designed for reduced diffraction (FIG. 9-14). If you find the cabinet work too demanding, you can install the same components in a rectangular box. Just make sure the cubic volume available for the two woofers is about the same as this, or two cubic feet. An enclosure with internal dimensions of about 10×16×26 inches before adding braces and wall linings should be about right.

Even if you build the enclosure as shown in the plans (FIG. 9-15), you can adopt easier construction methods if you feel the need. For example, in the enclosure shown in FIG. 9-14 the top panel was set into the sides by cutting the edges of the top and side panels at 45 degrees. You can instead cut the edges of the top and sides at 90 degrees and set the top above the sides. A hardwood trim could then cover the raw edges of the plywood top. If you do that, don't forget to shorten the sides by ¾ inch to retain the same internal dimensions.

It is desirable, but not mandatory, to flush-mount the drivers as was done in the cabinet shown in the photograph. If you do that, be sure to rout the recess in the speaker board that allows the driver to be flush mounted before cutting the 7½-inch holes. Otherwise, both

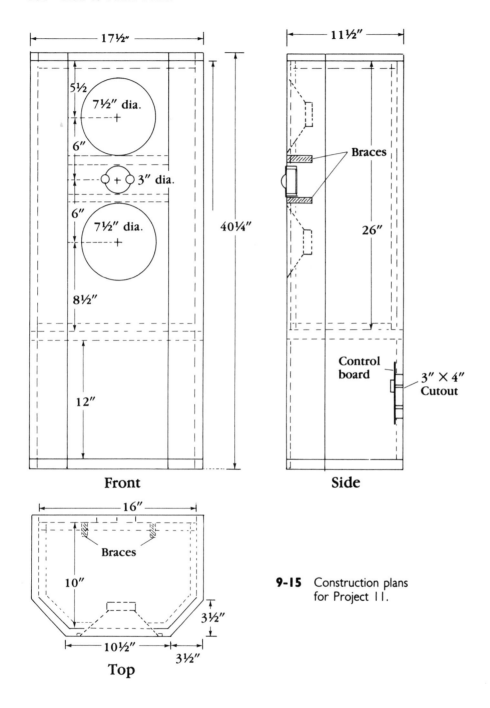

**Front**

**Side**

**Top**

**9-15** Construction plans for Project 11.

the pivot pin and the speaker board will have to be nailed down to hold them in place during the routing.

If you cut the angles for the joining panels at 23 degrees instead of exactly 22½ degrees, the fit at the outside corners will be tighter. Ideally the fit should be perfect all along the joint, but if you value appearance highly, use the 23-degree angle.

To make assembly easier, make up a jig to hold the side pieces while assembling the enclosure. Start with the front board and work to the back. That will permit you to block plane the bevels if that is needed to get a good fit on the angled cuts. Don't forget to glue in the partition when you put in the bottom piece. Make sure both those pieces are perpendicular to the other parts. That is necessary to get a good total fit.

No grille cloth was used on the speakers built for this project. If you want to hide the drivers, you can install a grille cloth on a light framework as shown in FIG. 9-16. Keep the framework as open as possible and install it on four 1-inch dowel spacers so that there is no interference with sound waves traveling horizontally from the drivers.

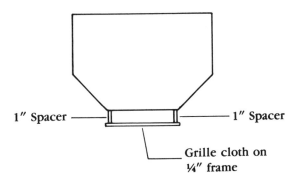

**9-16**  One way to attach a grille cloth to the cabinets of Project 11 without interfering with the low-diffraction design. When the grille cloth is spaced away from the front, as shown here, the sound can move horizontally from the drivers without encountering any obstructions or sharp corners.

This enclosure requires considerable bracing (FIG. 9-17). When completed, it should be lined with some kind of deadening material, as suggested in the parts list (TABLE 9-2). To follow the procedure used here and described in Chapter 4, use Liquid Nails to glue a layer

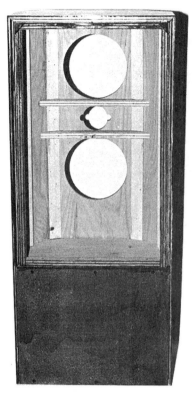

**9-17**  Internal construction of Project 11 enclosures.

of asphalt roofing material to the walls. Then cover that with a layer of thin, foam-backed carpet, gluing the foam layer to the asphalt material. Finally, install a layer of ½-inch carpet foam. After all of that, it's a good idea to cover the inside of the back panel with a 1- or 2-inch layer of polyester stuffing held in a nylon net.

The crossover network is diagrammed in FIG. 9-18. The only unusual feature there is the series connection for the 4-ohm woofers. The woofers used in this system are not noted for high power ability but offer good performance for a reasonable price. People who heard this system remarked that the sound was full bodied and "easy to take."

**Table 9-2  Parts List for Project 11: Double-Woofer Speakers**

| Pieces | Dimensions | Function |
|---|---|---|
| 1 | ¾ × 10½ × 40¼″ plywood | Speaker board |
| 2 | ¾ × 4¹⁵⁄₁₆ × 40¼″ plywood | Angled pieces |
| 2 | ¾ × 8 × 40¼″ plywood | Sides |
| 1 | ¾ × 16 × 26″ particle board | Back |
| 4 | 1 × 2 × 26″ hardwood | Corner cleats |
| 2 | 1 × 2 × 23″ hardwood | Braces on back |
| 2 | 1 × 2 × 12″ hardwood | Braces on speaker board |
| 10 ft. | ¾ × ¾″ hardwood | Cleats & glue blocks |
| 1 | ¾ × 12 × 16″ plywood or particle board | Back, lower compartment |
| 1 | ¼ × 4 × 6″ masonite | Control board |
| 9 ft.² | Asphalt roofing pieces | Wall liner |
| 9 ft.² | Thin, foam-backed carpet | Wall liner |
| 9 ft.² | ½″ carpet foam | Wall liner |
| | **Components** | |
| 2 | 8″ woofers, 4 ohm | Polydax HD20B25J4 |
| 1 | 1″ dome tweeter | Polydax HD25BAHR |
| 1 | 8″ L-pad, 15 W or better | Tweeter control |
| 1 | 4 uF capacitor, Mylar, 100 V | C1 |
| 1 | 12 uF capacitor, NP electrolytic, 100 V | C2 |
| 1 | 0.35 mH inductor, air core | L1 |
| 1 | 10 Ω resistor, 10 W | R1 |
| 1 | Speaker terminal plate | |

## PROJECT 12: Universal Aperiodic Enclosures

If you want to build your own speaker enclosures but don't want to fuss with crossover networks and other design problems, here is an enclosure project for you (FIG. 9-19). It is designed to be used with any good 6-×-9-inch coaxial or tri-axial car speakers. Such speakers usually have the midrange and tweeter drivers set out some distance from the speaker frame. To allow for that, this cabinet was designed with a thick grille frame (FIG. 9-20). If you plan to do this project, decide on a speaker first. You might be able to use a thicker speaker board along with a thinner grille frame to put the tweeters closer to the grille cloth.

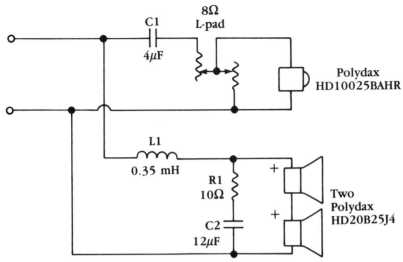

**9-18**   Diagram of Project 11 crossover network.

The cabinet shown in FIG. 9-20 was built by putting a solid walnut shell over a plywood inner box. If you prefer, you can use ¾-inch hardwood plywood for the entire enclosure. Just make sure the inner dimensions are the same as these. Even those can undergo minor changes, if necessary. But don't skimp on quality of construction.

## Construction Procedure

Except where a different assembly method is described, use wood glue and screws, or nails, for joining all parts. To duplicate the enclosure shown in FIG. 9-19, follow these steps.

**1**   Cut out the parts (TABLE 9-3).

**2**   Install a terminal plate on the back panel with about 18 inches of speaker cable wired and soldered to it. Seal the plate to the back with silicone rubber.

**3**   Assemble the bottom, duct rails, and duct top panel (FIG. 9-21).

**4**   Add inner box sides to base. Cut two small pieces of ⅜-inch plywood, about ¾×2¼ inches, to extend the inner sides to be flush with front edge of duct rails.

**5**   Install inner top.

**9-19** Project 12: Universal Aperiodic Enclosures.

**6** Install pine cleats and glue blocks inside inner box to reinforce corners and butt against speaker board and back (FIG. 9-22).

**7** Install back (FIG. 9-23).

**8** Caulk inner joints with silicone rubber sealer.

**9** Install walnut bottom, placing back edge flush with back of inner box, front edge extended 2⅛ inches forward from the front edge of inner box.

**10** Attach exterior sides (FIG. 9-24) and top. This can be done using glue alone if well clamped and allowed to stand overnight. Make sure top overhangs sides by ⅞ inch at front.

**9-20**   The grille frame for Project 12 must be thick enough to clear the front-mounted tweeters on coaxial or triaxial car speakers.

**11**   Prepare speaker board. Cut an oval hole to allow your speaker to be installed from outside the cabinet, which is about 6×8¾ inches for a typical 6×9 speaker. Locate hole center 7 inches from the top of the speaker board. Install T-nuts behind board to receive ³⁄₁₆-inch bolts for speaker mounting.

**12**   Install Acousta-Stuf. Carefully separate any clumped fibers to loosely fill box. Fill the duct also. This can take from 4 to 6 ounces, depending on how you do it. Feed the speaker cable out through the stuffing so it will be accessible.

**13**   Install speaker board with screws and silicone rubber sealer.

## Table 9-3   Parts List for Project 12: Universal Aperiodic Enclosures

| Pieces | Dimensions | Function |
|---|---|---|
| 2 | ⅜ × 10¼ × 19⅛" plywood | Sides, inner box |
| 1 | ⅜ × 7 × 9½" plywood | Duct top panel |
| 1 | ¾ × 10¼ × 20⅝" plywood | Back, inner box |
| 1 | ¾ × 10¼ × 11" plywood | Bottom, inner box |
| 1 | ¾ × 10¼ × 17⅝" plywood | Speaker board |
| 2 | ½ × 12⅞ × 19⅞" walnut | Sides |
| 1 | ½ × 13⁵⁄₁₆ × 13⅞" walnut | Bottom |
| 1 | ½ × 12¹⁄₁₆ × 13¾" walnut | Top |
| 2 | ¾ × 1¹⁄₁₆ × 19⅞" walnut | Grille trim, sides |
| 2 | ¾ × 1¹⁄₁₆ × 11¼" walnut | Grille trim, top & bottom |
| 6½ ft. | ⅜ × ¾" walnut molding | Trim |
| 10 ft. | ¾ × ¾" pine | Cleats & glue blocks |
| 2 | ¾ × 1½ × 7¾" pine | Duct rails |
| 2 | ¾ × ¾ × 18⅜" pine | Grille frame, sides |
| 2 | ¾ × ¾ × 10¼" pine | Grille frame, top & bottom |
| 1 | ¾ × ¾ × 8¾" pine | Grille frame, brace |
| 6 oz. | Acousta-Stuf or polyester fill | Damping material |

Components

| | | |
|---|---|---|
| 1 | 6 × 9" coaxial or triaxial speaker | Any good brand |
| 1 | Speaker terminal plate | |

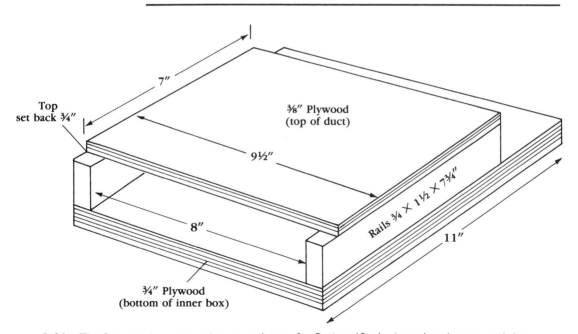

**9-21**   The first step in constructing an enclosure for Project 12: the inner box bottom and duct.

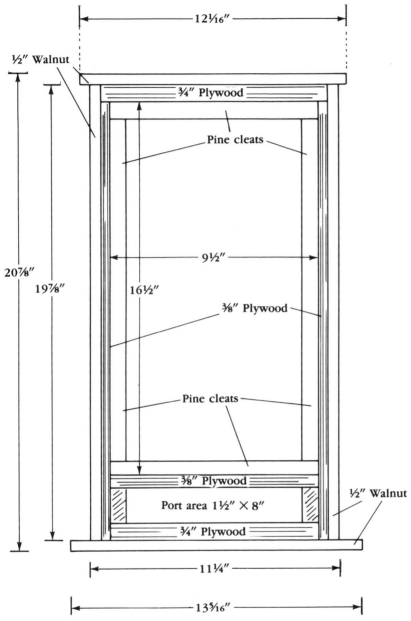

**9-22** Front view of cabinets in Project 12 during construction.

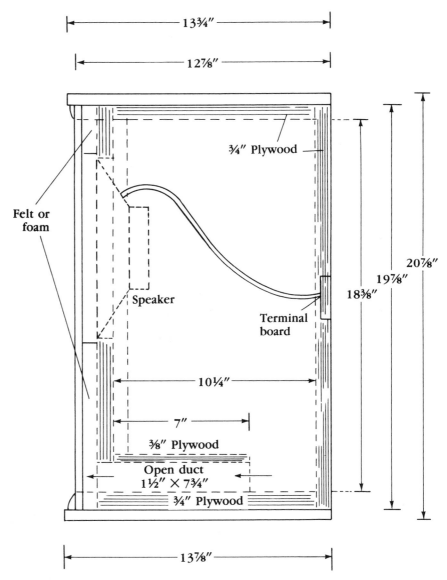

**9-23** Side view of Project 12 cabinets.

**14** Assemble grille frame. Position the cross brace so it doesn't interfere with the speaker or duct (FIG. 9-19).

**15** Cut and fit and trim that holds the grille frame in place to exact measurements of your cabinets. Cut the trim from ½-inch stock, 1¹⁄₁₆

**9-24**   Glue solid walnut sides to inner box in Project 12. For simpler construction, use ¾-inch hardwood plywood.

inches wide, mitered 45 degrees at corners (FIG. 9-25). This trim should measure about 11¼×19⅞ inches overall.

**16**   Cut out ⅜-inch molding ¾ inch wide and mitered at corners. The molding shown in the photograph (FIG. 9-20) was cut with a Sears ⅜-inch bead and quarter-round bit #25562.

**17**   Attach side molding, top and bottom, to the cabinet.

**18**   Attach front molding, top and bottom, to the grille trim frame. This permits the grille trim and front molding to be installed or removed in one piece (FIG. 9-26).

**19**   Sand cabinets and apply finish.

**20**   Install speakers with bolts.

**21**   Wire and install speakers with bolts and foam weatherstrip tape gaskets.

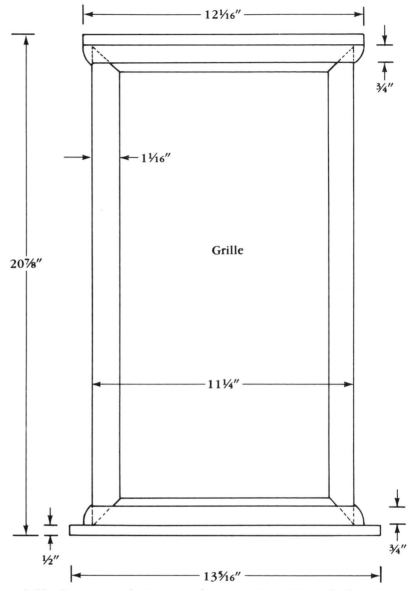

**9-25** Dimensions of grille trim — front and side molding — for Project 12.

**22** Cut and glue a layer of thick foam or felt to fill the space on the speaker board between the speaker and the grille frame.

**23** Test speakers. Make sure the polarity is the same on each speaker as described in Test 10 of Chapter 10. If bass seems over-

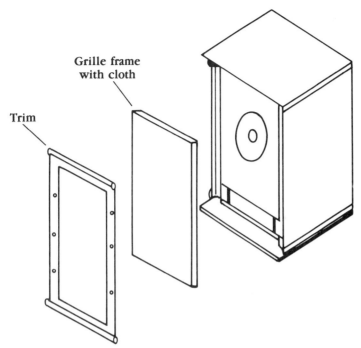

**Grille frame
with cloth**

**Trim**

**9-26**   This exploded view shows how grille frame and front trim
fit into cabinets of Project 12.

damped, remove some or all of the stuffing from the duct. If boomy, add more. Or, in either case, try closing the duct.

**24**   Cut and install grille cloth on grille frame. Choose a thin, open-weave cloth.

**25**   Place the grille frame in the cabinet and install the trim and molding assembly over it. Use three #4, 1-inch brass oval-head screws on each side, or, if you prefer, countersunk flathead screws.

Experiment with room position before making a final decision on where to place the speakers. Try raising the speakers by placing them on a temporary base. If you like the sound better there, you can build a set of stands for them.

$$Chapter \quad \textbf{10}$$

# Testing Speakers and Components

$E$ven when you can get reliable specifications on your speakers from the manufacturer, they will be only average values for a model run. To know the characteristics of the speakers you own, you must measure them yourself. And without tests, you are in the dark with an "orphan" speaker. To finely tune a system, tests are a great asset.

If you have no test equipment, you might be able to rent it. For that, read the tests carefully to be well prepared with any needed accessories before you obtain the electronic equipment.

## USEFUL TEST INSTRUMENTS

Here is a summary of the test instruments you are most likely to use.

### Digital VOM or Analog VTVM

A digital VOM (volt-ohmmeter) is a handy tool for many household tests such as checking continuity in appliances, measuring the voltage in your car batteries, or countless other uses. If you have no other equipment, you can use one to check the dc resistance of your speakers and by that, estimate their impedance.

You can use a digital VOM for some of the speaker tests in this chapter. It is particularly useful for low-frequency tests, but because of reduced high-frequency response, you must recalibrate the meter for every test as you go up in frequency. A digital VOM is good for accurately measuring the dc resistance of any speakers, especially small tweeters, because it sends a limited amount of current through the load (FIG. 10-1).

The old-fashioned VTVM (vacuum-tube voltmeter) is more convenient for high-frequency tests of drivers, but only if the VTVM is sensitive enough for the job. A special ac voltmeter (ACVM) can give full-scale deflection at low voltages such as 0.3 V, 0.1 V, or even lower. These meters have no facilities for measuring resistance or dc voltages.

## Audio Generator

An audio generator is another high-priority instrument (FIG. 10-2). Some cheap portable generators operate only at a predetermined set

**10-1**  A digital volt-ohmmeter (VOM) is useful for measuring dc resistance of voice coils, particularly that of small tweeters.

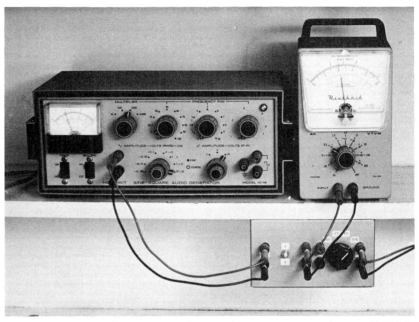

**10-2**  Typical setup for testing speakers: audio generator, ac voltmeter (ACVM), and instrument coupler.

of frequencies. They are adequate to see whether a driver is working or for very rough frequency tests, but they have very little usefulness for most of the tests in this book. Get a generator with continuously variable frequency controls.

### Sound Level Meter

Sound level meters are calibrated in decibels (dBs), so you can use them for frequency response tests. For accuracy, an SLM should be calibrated. Some models come with an average graph of the model's response curve. If you don't have an SLM, you can use a separate microphone and preamplifier combined with an ac voltmeter or oscilloscope to monitor response. The advantage of the SLM is that it is much more portable and convenient.

### Oscilloscope

Some of the tests require an oscilloscope, but a scope isn't necessary to design and build a speaker system, An oscilloscope just adds to the information you can get from your other tests, particularly information about phase relationships.

## Calculator or Computer

Any pocket calculator that has a log function is a great aid in testing and designing speaker systems. In fact, it is almost a necessity to use one for Keele's method of designing reflex speaker systems as described in Chapter 6.

If you have an IBM-compatible PC, you can use the computer program in Appendix B to quickly study the effect of design changes in the speaker systems you intend to build. You can for example, produce a frequency response graph of the theoretical performance of any woofer in any enclosure that you plan to use. Then you can explore how changes in cubic volume or tuning will alter that performance.

Beyond that, you can buy software that will help you design sophisticated crossover networks and even analyze phase response, mutual coupling of drivers with other drivers or ports, voice coil temperature at a given performance level, and many other aspects of speaker behavior. Such programs are an aid to speaker design, not an end. If you use any of them, be prepared to second guess the data they give you. Your ears should remain a part of your test equipment.

## Precision Resistor

To calibrate your equipment, you need at least one precision resistor. The most useful single value is 10 ohms. Ideally you should have a 10-ohm value plus another one that has a resistance about equal to that of most driver voice coils, or about 6 ohms. Resistors with 1 percent tolerance are good enough. After all, 1 percent of 10 is 0.1 ohm, and that is the limit of accuracy of most low-cost meters.

One way to obtain a super-accurate resistor is to purchase a 5- or 10-watt resistor of the value you need and take it to a local school laboratory to be measured. If the laboratory has a Wheatstone bridge, someone can probably measure the exact value. If you do that, write the value on the side of the resistor with a felt-tip pen.

## HOMEMADE TEST EQUIPMENT

You can make some test equipment. A standard test box, for example, can save time and probably gives more accurate data than other methods of measuring the equivalent volume of a speaker's suspension ($V_{AS}$). With some of today's speakers, the standard box doesn't work well because of the lack of a front gasket on the driver. For one of those, you can substitute the "Tupperware test," as described in Test 4.

To test speaker polarity and damping, you can make a simple tester from a battery, a resistor, and a switch.

## Standard Test Box

The ideal standard box technique would be to put the driver inside a box of known cubic volume and test it there. To do that you must remove a panel on the box each time you test a driver. Build such a box using T-nuts to receive the clamping bolts and a gasket to prevent air leaks. But it is more convenient to make an airtight box with no removable parts. The speaker can be tested *on* the box, not in it.

It is not necessary to make the box to any specific volume, but you must know the exact volume for calculations. And the cubic volume of the box should be a reasonable value for the driver. You could not get usable data for a 4-inch woofer if you tested it on a standard box built for a 15-inch woofer.

Figure 10-3 shows some standard box dimensions for testing woofers of various sizes. To build one of these boxes, follow this procedure:

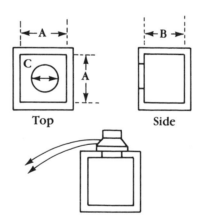

| Speaker size | Box dimensions | | | Box volume (ft.³) |
|---|---|---|---|---|
| Up to | A | B | C | |
| | | | To match | |
| 6″ | 8½″ | 5″ | speaker | 0.2 |
| 8″ | 10″ | 8⅝″ | 6¾″ | 0.5 |
| 10″ | 13″ | 10¼″ | 9″ | 1.0 |
| 12″ | 13″ | 10¼″ | 10½″ | 1.0 |

**10-3** Standard box dimensions for testing speakers of different sizes.

1 Cut out the parts.
2 Glue and nail the sides together.
3 Glue and nail down the speaker board.
4 Caulk the interior joints with silicone rubber sealer.
5 Glue and nail down the bottom panel.
6 Reach through the speaker hole to caulk the bottom joints.

To make a $V_{AS}$ test on a woofer with no front gasket, use any suitable container of circular crosssection that fits the woofer without interfering with the suspension. I call this the "Tupperware test" because I first used a Tupperware bowl designed to hold a head of lettuce in the refrigerator to measure the $V_{AS}$ of a driver with no gasket. Later, as more drivers were supplied with no gasket, I found that such items as plastic toy sand buckets, peanut butter jars, and oatmeal boxes also worked well (FIG. 10-4). To measure the cubic volume of such enclosures that are not simple geometric shapes, fill the container with water and measure the water with a graduated cylinder or other calibrated volumetric container.

## Polarity-Damper Tester

This little tester is very useful. To make it, get a tuna can or other chassis and the parts listed in TABLE 10-1. Then fit the components into the chassis and wire them according to the diagram in FIG. 10-5.

**10-4** A variety of round-mouthed containers can be used to test speakers that have no front gasket.

**Table 10-1   Parts List for Polarity-Damping Tester**

| Pieces | Description |
|---|---|
| 1 | SPDT mini toggle switch |
| 2 | Banana jacks |
| 2 | Banana plugs |
| 1 | Test probe wire, red |
| 1 | Test probe wire, black |
| 2 | Alligator clips, insulated |
| 1 | Resistor, 0.33 Ω |
| 1 | Battery holder for C battery, 1.5 V |
| 1 | Small chassis (i.e., tuna can) |

## Instrument Coupler

To avoid tangled webs of hookup wires, make up some kind of board to hold leads, switches, and precision resistors. One such coupler is shown in FIG. 10-2 and diagrammed in FIG. 10-6. It allows you to quickly switch from one load to another, or from a precision resistor to the speaker you are testing. You can alter the design to make the coupler more appropriate for the kinds of testing you will do. In building such a coupler, try to obtain a minitoggle switch with low contact resistance. When your coupler is built, measure it for internal resistance.

The simple coupler diagrammed in FIG. 10-6 is all that you need for most purposes, plus it includes constant-current impedance measurement capability.

## TESTS

Before testing a new speaker, it should be exercised for two hours. Set the audio generator at the speaker's free-air resonance frequency for the exercise period.

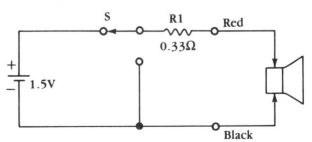

**10-5**   Schematic diagram of polarity-damping tester.

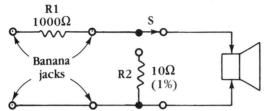

**10-6**   Schematic of simple coupler to connect test instruments and speaker.

### TEST 1: Free-Air Resonance (f,)

Hang the speaker in midair, if possible, with the driver axis horizontal. If the speaker is mounted on a baffle, the frequency of resonance will be lower by 1 or 2 percent.

### Method I: ACVM

Use the test setup shown in FIG. 10-7. For a woofer, set the audio generator to read about 200 Hz and choose a range on the ac voltmeter to give a reading near the lower end of the scale. Vary the frequency from the generator downward until you find the low frequency that produces the maximum reading on the voltmeter. Record this frequency as $f_s$.

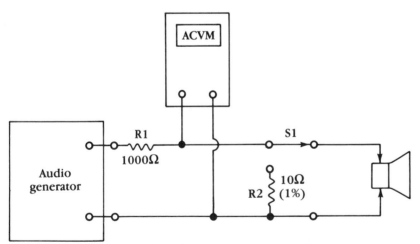

**10-7**   Test setup for voice coil impedance and other tests.

### Method II: Oscilloscope

Use the test setup shown in FIG. 10-8. Set the audio generator at 200 Hz. Sweep downward in frequency until the ellipse on the screen rotates to the left and closes into a straight line with maximum vertical deflection. Record this frequency as $f_s$.

Figure 10-9 shows the patterns you should see on your oscilloscope screen at frequencies above and below $f_s$.

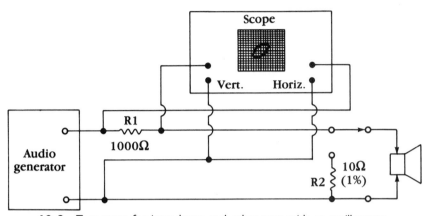

**10-8**  Test setup for impedance and other tests with an oscilloscope.

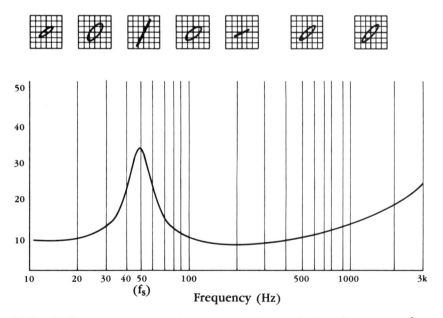

**10-9**  Oscilloscope patterns made at various points on the impedance curve of a speaker.

## TEST 2: Impedance

Use the test setup shown in FIG. 10-7.

### Method I: ACVM Test

Adjust the switch to put R2, the 10-ohm precision resistor, in the circuit. Adjust the generator output to read any 10 units on the ac voltmeter scale, but choose a point at the lower part of the scale. For example, with an 0.1-volt range, you might use 0.01 volt to represent a load of 10 ohms. That would be a good choice for a driver that has relatively wide variations in impedance at various frequencies. For one with narrower swings of impedance, you could choose the 0.03-volt range and use 0.01 volt to represent the load, but in this case that value would use one third of the scale. It would permit more accurate readings with an analog meter because of the greater space between unit marks.

Switch to the speaker and read the voltage at 10 Hz intervals from the lowest frequency of your audio generator to 100 Hz, at 100 Hz intervals to 1000 Hz (1 kHz), and at 1 kHz intervals at higher frequencies. Record the readings as ohms. The ACVM is reading volts, but the 1000-ohm series resistor converts the circuit to a constant-current one, making the voltage across the load proportional to the load's impedance.

In addition to recording the impedance at the stated intervals, record the frequency and impedance value of the load at $f_s$.

It might be useful to graph the impedance on log paper. For a 10 to 1000 graph, use two-cycle paper; for 10 to 10 kHz, three-cycle paper; and for a full audio band graph, four-cycle paper.

### Method II: Digital VOM Test

Use the test setup shown in FIG. 10-7, but with a digital VOM in place of the ACVM. To use a digital VOM, follow the procedure described for the ACVM with one difference. As you change the frequency interval upward, you must recalibrate the VOM. At each step upward in frequency, switch back to the 10-ohm precision resistor and increase the output of the audio generator to obtain a 10-unit reading on the VOM. Then switch to the speaker and make the reading at that frequency.

### Method III: Meraner Method

This is a refinement of Method II, suggested by David J. Meraner. It gives somewhat greater accuracy at high impedance readings and hence might be worthwhile for drivers with wide variations in impedance. Like Method II, it requires recalibration at each reading.

To use it, measure the precise value of R1, the series resistor. Switch to the speaker load. Connect the digital VOM across R1 and adjust the output of the audio generator to obtain a reading on the VOM of 0.001 times the exact value of R1. Then connect the VOM across the speaker load and make the reading. This reading, in millivolts, is proportional to speaker impedance.

## TEST 3: Speaker Q

This test requires two kinds of readings: the dc resistance of the driver's voice coil and its impedance.

### Method I: Small's Test

Calibrate your ohmmeter with a 5- to 10-ohm precision resistor. Measure the dc resistance of the voice coil. Record this value as $R_E$.

Set the audio generator to the frequency of $f_s$. Read the value of the driver's impedance at that frequency and record it as $Z_{MAX}$. Calculate the value of $r_o$ from:

$$r_o = Z_{MAX}/R_E$$

Record this value. Find $\sqrt{r_o}$ and record its value. Calculate the value of reduced impedance (Z') where:

$$Z' = \sqrt{r_o}R_E$$

Find the frequencies above and below $f_s$ where driver impedance equals Z'. Record these frequencies as $f_1$ and $f_2$.

Check the accuracy of your work by this test:

$$f_s = \sqrt{f_1 f_2}$$

The solution to this formula should be accurate within about 1 Hz or 2 percent, whichever is greater.

Find the driver's mechanical Q ($Q_{MS}$) by:

$$Q_{MS} = \frac{f_s \sqrt{r_o}}{f_2 - f_1}$$

Find the driver's electrical Q ($Q_{ES}$) by:

$$Q_{ES} = Q_{MS}/r_o - 1$$

Then total Q ($Q_{TS}$) is:

$$Q_{TS} = \frac{Q_{MS}Q_{ES}}{Q_{MS} + Q_{ES}}$$

## Method II: Quick Q Test

Measure $R_E$, $f_s$, and $Z_{MAX}$. Find $f_1$ and $f_2$ where $Z' = 0.707\ Z_{MAX}$. Then:

$$Q = \left(\frac{f_s}{f_2 - f_1}\right)\left(\frac{R_E}{Z_{MAX}}\right)$$

## TEST 4: Finding V$_{AS}$

Use the test setup shown in FIG. 10-7.

## Method I: Standard Box

For dimensions of a standard box for your speaker, see FIG. 10-3. Press the speaker down over the cutout in the standard box (FIG. 10-10). Apply enough pressure to make a good seal. Measure the new frequency of resonance with the speaker on the box. Record this frequency as $f_{ct}$. Find $V_{AS}$ by:

$$V_{AS} = 1.15\ [(f_{ct}/f_s)^2 - 1]\ V_B$$

where $V_B$ is the cubic volume of the standard box.

## Method II: Tupperware Test

This test is useful for drivers that have no front gasket. Find a container, such as a Tupperware bowl, that has a circular mouth to fit the driver. If necessary, line the rim of the container with foam weatherstrip tape. Press the driver down on the empty container and record $f_{ct}$. Remove the driver and measure the cubic volume of the test container by filling it with water from a calibrated vessel. If the vessel is calibrated in liters, use this conversion: 1 cubic foot = 28 liters.

Find $V_{AS}$ by the formula shown for Method I.

## Method III: Ported-Box Test

Use the test setup shown in FIG. 10-7. Install the driver in any ported box that has no damping material in it. If the port is adjustable, tune the box approximately to $f_s$, but this step is not mandatory.

**10-10**   Press the speaker down on the standard box to make the test for V$_{AS}$.

Set the audio generator to the lowest frequency range. Sweep the range upward to find the three critical frequencies of a ported system: the lower impedance peak, f$_L$; the low point above that peak, which occurs at the box tuning frequency, f$_B$; and the upper impedance peak, f$_H$.

Calculate V$_{AS}$ from:

$$V_{AS} = V_B \frac{(f_H + f_B)(f_H - f_B)(f_B + f_L)(f_B - f_L)}{f_H^2 f_L^2}$$

This method has the advantage that minor leaks will not affect the results. A carefully executed standard box test is just about as accurate with most speakers.

## TEST 5: Tuned Frequency of a Ported Box (f$_B$)

If you use more than one of the methods described here and get inconsistent data, try to correct any enclosure or driver flaws that can cause such variations.

## Method I: Impedance Method

Use the test setup shown in FIG. 10-7. Set the audio generator to the lowest frequency range. Sweep the range upward to find the three critical frequencies of a ported system: the lower impedance peak, $f_L$; the low point above that which is approximately at $f_B$; and the upper impedance peak, $f_H$. Record these frequencies.

Close the port with an airtight seal. Locate the frequency of the system resonance, which occurs at an easily identifiable single impedance peak. Record this frequency as $f_C$. Calculate $f_B$ by:

$$f_B = \sqrt{f_L{}^2 + f_H{}^2 - f_C{}^2}$$

## Method II: Impedance and Phase Method

Use the test setup shown in FIG. 10-8. Set the audio generator to the lowest frequency range. Sweep upward to find the same three critical frequencies mentioned above. Locate the exact frequency of $f_B$ by noting the point where the shrinking ellipse closes to a straight line, indicating a zero phase condition. Record this frequency as $f_B$.

If the $f_B$ obtained by this method differs from that obtained by Method I, it could indicate unusual leakage in the system. Check the driver surround and dust cover. If the driver is a cheap one, consider treating the leaky part with a thin coat of silicone rubber.

## Method III: Visual Test

Set the speaker on its back. Dust some powdered chalk on the driver cone. Adjust the output of the audio generator to provide a 5-volt signal to the driver. Adjust a bright light so you can watch the cone vibration while you sweep the frequency range near $f_B$. At $f_B$, the cone will appear to stand still.

If there are significant leaks in the box, the cone will vibrate at all frequencies. Search out and stop all leaks until there is no apparent movement at $f_B$.

## Method IV: SLM or Listening Test

Use the test setup shown in FIG. 10-11. Note that this test can be used even though you don't have the driver that will be installed in the box. Drill a ¼-inch hole in the box, preferably in an area that will be cut out for the driver. Set the audio generator to the lowest range and sweep the frequency range near the expected frequency of $f_B$. At $f_B$, the SLM or your ears, should detect a peak in sound output from the port (FIG. 10-12).

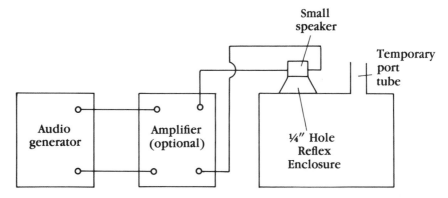

**10-11**   Instrument setup for making Test 5 (Tuned Frequency of a Ported Box), Method IV.

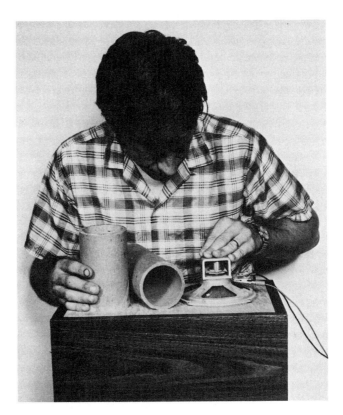

**10-12**   Tuning a ported box by Test 5, Method IV.

## TEST 6: Cabinet Vibration Test

Use the test setup shown in FIG. 10-13 with the driver installed in the cabinet. The SLM is not required.

### Method I: Fingertip Test

Have someone operate the audio generator to sweep the bass and lower midrange frequencies while you run your fingertips over each panel of the enclosure to enable you to detect the areas on each panel that need further bracing.

### Method II: Chalk Test

Lay the cabinet on the floor with the worst offending panel, as discovered in Method I, facing upward. Sprinkle some powdered chalk on the panel and repeat the test. The area of greatest vibration will have less chalk remaining after this test.

## TEST 7: Frequency Response Test

Use the test setup shown in FIG. 10-13. If you live in a quiet neighborhood, do this test outdoors on a calm day. Face the speaker into a large open space with no reflecting surfaces near it. Set the speaker on a stand about 1 to 2 feet above the ground. Place the SLM a foot or two from the apparent acoustic center of the speaker.

Set your audio generator at 1000 Hz and adjust the output level to get a useful deflection on the SLM, such as 0 dB. Run up and down the frequency scale and record the frequencies and amplitudes of any peaks or valleys in the speaker's response. Repeat the test after changing the position of the SLM in respect to the speaker, both farther and nearer.

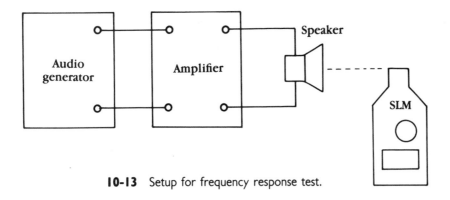

**10-13**  Setup for frequency response test.

If you must do this test indoors, take several readings at various parts of the room. This test can be useful to help design notch filters but should be used only if your ears agree with the test.

## TEST 8: Peak Displacement Volume Cone Test

Use the test setup shown in FIG. 10-13 minus the SLM. The driver is unbaffled. Hold the speaker to your ear and listen to the near field output while sweeping the 25 to 60 Hz range with a constant drive voltage to speaker. Increase drive voltage until you hear a sudden increase in distortion, indicated by a change in tone at a constant frequency. Estimate the linear displacement of the cone at this point. You can make this estimate by noticing the ghost image of the cone's peak displacement. Record this distance as $X_{MAX}$. Calculate the peak displacement volume by:

$$V_D = AX_{MAX}$$

where A is the effective cone area. You can estimate effective cone area by measuring the diameter of the radius (r) of the inner piston area plus half the width of the suspension. Then

$$A = \pi r^2$$

## TEST 9: FM Hiss Test

Test setup includes an FM receiver and speakers. Switch the receiver to mono mode. Tune in a strong station. Sit in your listening chair with speakers appropriately located. Check to see if the apparent sound source is midway between the two speakers. If it is not, balance the output until it appears there.

Remove the antenna from the FM receiver and detune to obtain a pure hiss between stations. Return to your chair and see if the hiss appears to come from the same spot as the program from the station. If not, adjust the tweeter controls on the speakers for symmetrical high-frequency response. Connect the antenna and try the station signal again. If the sound is too bright or too dull, readjust the tweeter controls and redo the hiss test.

The hiss test is one way to detect speaker coloration. If the sound is that of one blowing into cupped hands, the speakers' sound is more or less colored. This test can be used to adjust stuffing in a speaker enclosure.

## TEST 10: Polarity Test

With careful listening, you can probably omit this test. But when you consider that many audio showrooms have speakers connected with reversed polarity, maybe everyone should do the test.

### Method I: Homemade Tester Method

Use the homemade polarity-damping tester diagrammed in FIG. 10-5. Connect the leads of the tester to the speaker terminals. Throw the switch, S, and observe the cone movement. Reverse the leads as necessary to make the speaker cone move outward at the circuit "on" connection and inward at the "off" position. Then mark the speaker terminal that has the red lead is connected to it with a red dot or plus sign.

### Method II: Newspaper Test

If you have a commercial speaker with a non-removable grille cloth, place a single sheet of newspaper over the position of the woofer. Operate the tester used in Method I and observe the movement of the paper. Mark the terminals as for Method I.

### Method III: Listening Test

Test setup includes speakers and a receiver. Place the two speakers close together, face to face. Choose a music program with heavy bass. Listen for a few minutes, then shut off the receiver and reverse the leads to one speaker. Listen again. Choose the connection that produces more bass response.

## TEST 11: Voice Coil Inductance Test ($L_E$)

Most of the prescribed tests for voice coil inductance are unreliable. *Theoretically*, you can measure the impedance at a high frequency and subtract the value of $R_E$ *as a vector* from the total impedance to get the inductive reactance, $X_L$. This does not seem to give very good accuracy. Instead, the tests here are suggested for an approximation of the voice coil inductance.

### Method I: Rule-of-Thumb Test

Measure the dc resistance of the voice coil and record the value as $R_E$. Then use the test setup shown in FIG. 10-7. Sweep upward in frequency until you find the frequency where the impedance is equal to $2R_E$. Record this as $f_D$. Calculate $L_E$ by:

$$L_E = 0.3R_E/f_D$$

## Method II: Oscilloscope Test

Use the test setup shown in FIG. 10-8. Switch the speaker into the test circuit. Set the audio generator frequency control to $f_s$, then sweep upward until the ellipse on the screen closes into a straight line. This is the minimum impedance level, where the speaker's capacitive reactance equals its inductive reactance. Adjust the scope controls for equal vertical and horizontal deflection so that the line is inclined at 45 degrees and centered on the screen.

Switch the precision resistor into the test circuit. Adjust the output of the audio generator until the vertical deflection is 1 centimeter for a 10-ohm resistor.

Switch to the speaker. Measure the impedance value from the total vertical deflection on the grid (FIG. 10-14). Record this value as Z. Measure the sine of the phase angle by:

$$\sin \theta = A/B$$

The reactive component of the impedance is:

$$X_L = \sin \theta \, Z$$

But if you calibrated the oscilloscope with the 10-ohm resistor, there is no need to do any arithmetic with the sine function. The reactive component of the impedance in ohms is the reading A. You can find $L_E$ by:

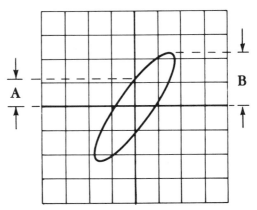

**10-14** Oscilloscope impedance and inductive reactance measurement.

$$L_E = X_L / (2\pi f)$$

where f is the test frequency, 1 kHz in this case.

## TEST 12: Inductance of a Crossover Coil

The oscilloscope test shows resonance more sharply than the ACVM test, but either test is useful. Resonance tests can also be done with a coil of known value to identify an unknown capacitor.

### Method I: Resonance Test With ACVM

Use the test setup shown in FIG. 10-7, but substitute the unknown coil, wired in parallel with a capacitor of known value, for the speaker in FIG. 10-7. Use a Mylar or close-tolerance polypropylene capacitor, if possible.

Switch to the various frequency bands on your audio generator until you find the range where the voltage across the coil and capacitor rises sharply. Locate the exact frequency of maximum voltage. Record the frequency as f. Calculate the inductance of the coil by:

$$L = 0.025/f^2C$$

### Method II: Resonance Test with Oscilloscope

Use the test setup shown in FIG. 10-8, but substitute the unknown coil, wired in parallel with a capacitor of known value, for the speaker in FIG. 10-8.

Follow the same procedure summarized in Method I. The frequency of maximum circuit impedance will be indicated by a sharp increase in vertical height of the ellipse on the oscilloscope. At the precise frequency of resonance, the ellipse will close into a straight line. Record that frequency as f and use the formula above to find L.

### Method III: Simple Impedance Test

Measure the dc resistance of the unknown coil and record the value as R. If the resistance is below 1 ohm, it can be disregarded in most cases.

Use the test setup shown in FIG. 10-7, but substitute the unknown coil for the speaker. Follow the procedure for running an impedance test for a speaker, but only one test reading is mandatory. Set the audio generator frequency control at 1000 Hz, 1 kHz, and read the impedance of the coil (Z) at that frequency (f). Calculate the inductive reactance of the coil by:

$$X_L = \sqrt{Z^2 - R^2}$$

As mentioned previously, if R is less than 1 ohm, ignore it. Calculate coil inductance by:

$$L = X_L/(2\pi f)$$

### TEST 13: Filter Components Resonance Test

Use the test setup shown in FIG. 10-15. Select R to equal the impedance of the speaker. Sweep the frequency range near the expected resonance frequency of the circuit. Note the frequency where the voltage across the inductor and capacitor falls to a minimum. This is the crossover frequency for a network made with these parts. This test can be used for any circuit for which it is desirable to make the inductive reactance equal the capacitive reactance at a designated frequency for a given speaker.

### TEST 14: Crossover Network/Speaker Response Test

Use the test setup shown in FIG. 10-16. When testing the woofer leg of the circuit, disconnect the tweeter and insert a resistor that is equal in resistance to the tweeter's impedance in the tweeter branch. When testing the high-frequency branch, use a resistor in the woofer circuit. The drivers should be installed in the enclosure with the crossover network outside so that changes can be made in crossover components if necessary.

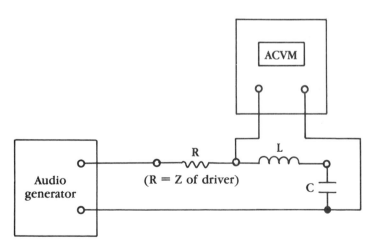

**10-15** Setup for Test 13: Filter Components Resonance Test.

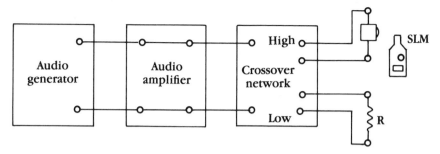

**10-16**  Setup for Test 14: Crossover Network/Speaker Response Test.

## Method I: SLM Test

Place the SLM on the axis of the driver to be tested. Sweep the frequency range near the crossover point to see if the crossover network is functioning.

In the case of a first-order network, you can get an estimate of the effectiveness of the inductor at any frequency by setting the output for a low SLM reading and then shorting the capacitor or coil under test. Observe the change in sound level by the SLM.

After testing the woofer branch, disconnect the woofer and substitute a resistor for it while you test the tweeter circuit.

## Method II: Listening

This involves nothing more than disconnecting the drivers, one at a time as in Method I, and simple listening to the woofer alone or the tweeter alone. This kind of test takes some experience, but after doing it a few times, you might be able to isolate problems.

## TEST 15: Speaker Damping Test

Use the homemade tester diagrammed in FIG. 10-5. Connect the leads of the tester to the speaker terminals with the driver installed in its enclosure. Flip the switch on and off. There will always be a slight difference in the two actions, but the ideal is to hear a "click" instead of a "bong." Note that when the tester switches on, the sound is more like a click than when it switches off. If the difference is significant, you might need to add mechanical damping to the driver by stapling a fiberglass blanket over the back of it. The blanket should be stretched tightly. Make the final adjustment by doing a listening test.

## TEST 16: Notch Filter Tests

Use the test setup shown in FIG. 10-7, but connect the notch filter in place of the speaker. Follow the procedure for an impedance test on a speaker. Find the frequency of highest impedance and record that as f. Read the value of maximum impedance and record it as Z. Calculate Q by this formula:

$$Q = Z2\pi fC$$

Or, measure Q directly by recording f and then finding the two frequencies, one below resonance $(f_1)$ and the other above resonance $(f_2)$ where the impedance falls to 0.707 that of the value at resonance. Calculate Q by:

$$Q = f/\Delta f$$

where f is the resonance frequency and $\Delta f$ is $f_1 - f_2$. Check the accuracy of your work by:

$$f = \sqrt{f_1 f_2}$$

## TEST 17: Critical Listening Tests

Test setup is with speakers connected to a receiver or amplifier. If you have facilities for a recording, make a tape of someone speaking, preferably someone you know well. You can make minor changes in crossover networks or in tweeter adjustments and choose the more natural combination.

If you must rely on commercial recordings, again use voice for at least part of your testing. Voice is good for setting tweeter level. If the speaker or singer sound like they have a lisp, you have the tweeter control too high.

Beware of comparing two speakers in a room. Room position can overcome many other minor differences. If you live in a quiet area, take them outdoors for comparison.

When operated in a mono mode, two stereo speakers should form a sound image midway between the speakers. If they don't, something needs adjusting. Attempt to remedy that situation using Test 9, the FM hiss test.

With a good piano recording, you should be able to hear the hammer tone. Drums should appear taut, not thuddy. And the stereo image should be stable. If it isn't, you might need to check the values of the parts in your crossover networks. Remember the importance of symmetry for good stereo performance.

In some ways, this is the most important test of all. Learn to listen critically, and you will get more out of all your listening.

*Appendix* **A**

# Mail-Order
# Houses

A & S SPEAKERS
3170 23rd St.
San Francisco, CA 94110
(415) 641-4573

AUDIO CONCEPTS
1631 Caledonia St.
P.O. Box 212
La Crosse, WI 54602-0212
(800) 346-9183 for orders or a catalog
(608) 784-4570 for product information

CRUTCHFIELD
1 Crutchfield Park
Charlottesville, VA 22906
(800) 336-5566 for a catalog
(800) 446-1640 for information or orders

MADISOUND
8608 University Green
Box 4283
Madison, WI 53711
(608) 831-3433
(608) 836-9473 for Audio bulletin board
(300/1200/2400/9600 MNP 5 USR HST)

McGEE RADIO
1901 McGee St.
Kansas City, MO 64108-1891
(816) 842-5092
(800) 876-2433 for credit card orders

MAHOGANY SOUND
2430 Schillingers Rd. #488
Mobile, AL 36695
(205) 633-2054 (after 5 p.m. CST)
Sells Acousta-Stuf

MCM ELECTRONICS
650 Congress Park Dr.
Centerville, OH 45459-4072
(513) 434-0031
(800) 543-4330 for orders or a catalog
(800) 762-4315 in Ohio
(800) 858-1849 in Hawaii or Alaska
(800) 829-9491 in Canada

MENISCUS
2442 28$^{TH}$ St. SW Ste. D
Wyoming, MI 49509−2158
(616) 534-9121

OLD COLONY SOUND LAB
P.O. Box 243
Peterborough, NH 03458
(603) 924-6371
Computer software for speaker design.

PARTS EXPRESS
340 E. First St.
Dayton, OH 45402
(513) 222-0173
(800) 338-0531 for orders or a catalog

SRS ENTERPRISES
318 South Wahsatch Ave.
Colorado Springs, CO 80903
(719) 475-2545

WOODWORKER'S SUPPLY OF NEW MEXICO
5604 Alameda NE
Albuquerque, NM 87119-2119
(505) 821-0578
Tools and hardware

ZALYTRON
469 Jericho Turnpike
Mineola, NY 11501
(516) 747-3515

# Audio Magazines

*Audio*
P.O. Box 51011
Boulder, CO 80321-1011
Articles, equipment tests and reports,
record reviews, and a few projects.

*Audio Amateur*
P.O. Box 576
Peterborough, NH 03458
(603) 924-9464
Wide range of audio projects, including
modifications for upgrading components.

*The Audio Critic*
P.O. Box 978
Quakertown, PA 18951
Equipment tests and evaluations.

*Popular Electronics*
(combined with *Hands-on Electronics*)
P.O. Box 338
Mt. Morris, IL 61054-9932
Wide range of electronic projects.

*Radio-Electronics*
P.O. Box 51866
Boulder, CO 80321-1866
Audio, video, shortwave, and computers—
with projects.

*Speaker Builder*
P.O. Box 494
Peterborough, NH 03458
(603) 924-9464
State-of-the-art speaker projects.

*Stereo Review*
P.O. Box 53033
Boulder, CO 80321-2033
Articles, equipment tests, and record reviews.

*Stereophile*
P.O. Box 364
Mt. Morris, IL 61054
Subjective evaluation of "high-end" audio
equipment and record reviews.

*Voice Coil*
P.O. Box 176
Peterborough, NH 03458
Loudspeaker industry news, new patents,
and technical reports.

*Appendix* **C**

# Loudspeaker Design Program

$T$he following program was written by Hal Finnell and is based on an earlier program by Robert L. Caudle. To obtain graphics from the program (graphs showing plots of the design you've chosen), you must either have CGA or use an adapter program. One such program, known as HGCIBM, is available from J. Gary Batson of Athena Digital, 506 Walker St., Opelika, AL 36801.

To use the following program, follow these basic steps.

## To Write and Save a BASIC Program

**1** Boot the system and run either GWBASIC or BASICA for your system.

**2** Type in the program.

**3** To save, as the last line type SAVE" (usually F4) and the name of the program, "LDP."

**4** Exit BASIC.

## To Run the Program Saved on a BASIC Disc

**1** Type GWBASIC or BASICA and the name of the program (LDP). Return.

**2** Turn CAPS LOCK on.

## To Run the Program If on a Second Disc

**1**  Boot the system and run GWBASIC or BASICA.

**2**  Remove BASIC disc; insert LDP disc.

**3**  Turn CAPS LOCK on.

**4**  Type LOAD" (usually F3) and the name of the program (LDP). Return.

**5**  Type RUN (usually F2). Return.

```
10 REM --------- LOUDSPEAKER DESIGN PROGRAMS --------------------
15 REM --------- WRITTEN BY ROBERT L. CAUDLE --------------------
20 REM --------- BASED ON THE EQUATIONS OF ---------------------
25 REM --------- D.B. KEELE AND RICHARD SMALL ------------------
30 DIM M(200):SCREEN 2
40 CLS:PRINT TAB(15)"LOUDSPEAKER SYSTEM DESIGN PROGRAMS":PRINT:PRINT
50 PRINT"1.   DRIVER PARAMETERS"
60 PRINT"2.   VENTED BOX DESIGN"
70 PRINT"3.   CLOSED BOX DESIGN"
80 PRINT"4.   END"
86 X=6.7:Y=3
90 PRINT:INPUT"SELECT 1 - 4";P
100 IF P = 1 THEN 140
110 IF P = 2 THEN 530
120 IF P = 3 THEN 1370
130 IF P = 4 THEN END
140 CLS:PRINT TAB(23)"DRIVER PARAMETERS":PRINT:PRINT
150 L$ = "N":R$ = "N"
160 INPUT"ENTER DRIVER NAME";D$
170 INPUT"ENTER D.C. RESISTANCE OF VOICE COIL";RE
180 INPUT"ENTER FREE-AIR RESONANCE"; FS
190 INPUT"ENTER IMPEDANCE AT FREE-AIR RESONANCE";ZMAX
200 RO = ZMAX/RE
210 RF = SQR(RO)*RE
220 PRINT"ENTER FREQ. BELOW FREE-AIR =";RF;" OHMS";
230 INPUT F1
240 PRINT"ENTER FREQ.ABOVE FREE-AIR =";RF;" OHMS";
250 INPUT F2
260 QMS = FS*SQR(RO)/(F2-F1)
270 QES = QMS/(RO-1)
280 QTS = QMS*QES/(QMS + QES)
290 INPUT"ENTER TEST BOX VOLUME";TVB
300 INPUT"ENTER DRIVER RESONANCE IN TEST BOX";TFS
310 VAS = TVB*(1.149*((TFS/FS)^2-1))
320 CLS:PRINT TAB(23)"DRIVER PARAMETERS":PRINT:PRINT
330 PRINT D$
340 PRINT"VOICE COIL RESISTANCE(RE) =";RE;" OHMS"
350 PRINT"FREE-AIR RESONANCE(FS) =";FS;" HZ"
360 PRINT"QMS =";QMS
370 PRINT"QES =";QES
380 PRINT"QTS =";QTS
390 PRINT"VAS ="VAS
400 PRINT:PRINT:INPUT"LINE PRINT OUTPUT (Y) ";L$
```

```
410 IF L$ <> "Y" THEN 500
420 LPRINT TAB(23)"DRIVER PARAMETERS"
430 LPRINT:LPRINT:LPRINT:LPRINT D$
440 LPRINT"VOICE COIL RESISTANCE (RE) =";RE;RE;" OHMS"
450 LPRINT"FREE-AIR RESONANCE (FS) =";FS;" HZ"
460 LPRINT"QMS =";QMS
470 LPRINT"QES =";QES
480 LPRINT"QTS =";QTS
490 LPRINT"VAS =";VAS
500 INPUT"ANOTHER DRIVER (Y)";R$
510 IF R$ ="Y" THEN 140
520 GOTO 40
530 CLS:PRINT TAB(23)"VENTED BOX DESIGN":PRINT:PRINT
540 L$ ="N"
550 INPUT"DRIVER NAME ";D$
560 INPUT"ENTER QTS";QTS
570 INPUT"ENTER VAS";VAS
580 INPUT"ENTER FS";FS
590 VB = 15*QTS^2.87*VAS
600 FB = .42*QTS^-.9*FS
610 FH = .26*QTS^-1.4*FS
620 CLS:R$ = "N"
630 PRINT TAB(25)"B4-ALIGNMENT":PRINT:PRINT
640 PRINT"VB = "; VB
650 PRINT"FB = ";FB;" HZ"
660 PRINT"F3 = ";FH;" HZ"
670 H = 0
680 PRINT:PRINT:INPUT"CHANGE BOX SIZE (Y)";R$
690 IF R$ <> "Y" THEN 740
700 PRINT:INPUT"ENTER NEW VB ";VB
710 FB  = FS*(VAS/VB)^.32
720 FH = FS*SQR(VAS/VB)
730 H = 20*LOG(2.6*QTS*(VAS/VB)^.35)/LOG(10)
740 CLS:PRINT TAB(23)"VENTED BOX DESIGN":PRINT:PRINT
750 B$ = "VENT" : R$ = "N" : L$ = "N"
760 PRINT D$
770 PRINT"VB = ";VB
780 PRINT"FB = ";FB;" HZ"
790 PRINT"F3 = ";FH;" HZ"
800 PRINT"PEAK OR DIP IN RESPONSE = ";H;" DB"
810 PRINT:INPUT"CHANGE BOX SIZE (Y)";R$
820 IF R$ = "Y" THEN 700
830 INPUT"LINE PRINT OUTPUT (Y)";L$
840 IF L$ <> "Y" THEN 900
850 LPRINT TAB(23)"VENTED BOX DESIGN":LPRINT:LPRINT
855 LPRINT D$
860 LPRINT"VB = ";VB
870 LPRINT"FB = ";FB;" HZ"
880 LPRINT"F3 = ";FH;" HZ"
890 LPRINT"PEAK OF DIP IN RESPONSE = ";H;" DB"
900 INPUT"VENTED BOX RESPONSE GRAPH (Y)";R$
910 IF R$ = "Y" THEN 1020
920 INPUT"CUSTOM DESIGN (Y)";R$
930 IF R$ <> "Y" THEN 1340
940 CLS:PRINT TAB(20)"CUSTOM VENTED BOX DESIGN"
950 B$ = "":R$ = "N"
960 PRINT:PRINT:INPUT"CHANGE VB (Y)";R$
970 IF R$ <> "Y" THEN 985
980 INPUT"ENTER NEW VB ";VB
```

```
985 R$ = "N"
990 INPUT"CHANGE BOX TUNING (Y)";R$
1000 IF R$ <> "Y" THEN 1020
1010 INPUT"ENTER NEW FB ";FB
1020 CLS:PRINT TAB(15)"GRAPHICS CALCULATION IN PROGRESS"
1030 A = (FB^2)/(FS^2)
1040 B = A/QTS+(FB/(7*FS))
1050 C = 1+A+(FB/(7*FS*QTS))+(VAS/VB)
1060 D = 1/QTS+(FB/(7*FS))
1070 FOR F = 20 TO 200 STEP 5
1080 F9 = F/FS:F5 = F9^2
1090 F4 = F9^4:F3 = F9^3
1100 F6 =(F4-C*F5+A)^2
1110 F7 = (B*F9-D*F3)^2
1120 M(F) = 20*(LOG(F4/(F6+F7)^.5)/LOG(10))
1130 NEXT
1140 GOSUB 1900
1150 IF B$ = "VENT" THEN 740
1160 CLS:PRINT TAB(20)"CUSTOM VENTED BOX DESIGN"
1165 R$ = "N"
1170 PRINT:PRINT:PRINT D$
1180 PRINT"VB = ";VB
1190 PRINT"FB = ";FB;" HZ
1200 PRINT"QTS = ";QTS
1210 PRINT"FS = ";FS;" HZ"
1220 PRINT"VAS = ";VAS
1230 PRINT:PRINT:INPUT"CHANGE VB OR FB (Y)";R$
1240 IF R$ = "Y" THEN 940
1250 INPUT"LINE PRINT OUTPUT (Y)";L$
1260 IF L$ <> "Y" THEN 1340
1270 LPRINT TAB(20)"CUSTOM VENTED BOX DESIGN":LPRINT:LPRINT
1280 LPRINT D$
1290 LPRINT"VB = ";VB
1300 LPRINT"FB = ";FB;" HZ"
1310 LPRINT"QTS = ";QTS
1320 LPRINT"FS = ";FS;" HZ"
1330 LPRINT"VAS = ";VAS
1340 INPUT"ANOTHER VENTED BOX DESIGN (Y)";R$
1350 IF R$ = "Y" THEN 530
1360 GOTO 40
1370 CLS:PRINT TAB(23)"CLOSED BOX DESIGN":PRINT:PRINT
1380 L$ = "N":R$ = "N"
1390 INPUT"DRIVER NAME";D$
1400 INPUT"ENTER QTS";QS
1410 INPUT"ENTER VAS";VAS
1420 INPUT"ENTER FS";FS
1430 INPUT"ENTER VB";VB
1440 A = VAS/VB
1450 FC = FS*SQR(A+1)
1460 QTC = (FC*QS)/FS
1470 F3 = FC*SQR(((1/QTC^2-2)+SQR((1/QTC^2-2)^2+4))/2)
1480 CLS:PRINT TAB(23)"CLOSED BOX DESIGN"
1485 R$ = "N"
1490 PRINT:PRINT:PRINT D$
1500 PRINT"F3 = ";F3;" HZ"
1510 PRINT"QTC = ";QTC
1520 PRINT"VB = ";VB
1530 PRINT:INPUT"CHANGE BOX SIZE (Y)";R$
1540 IF R$ <> "Y" THEN 1570
```

```
1550 PRINT:INPUT"NEW VB ";VB
1560 GOTO 1440
1570 CLS:PRINT TAB(23)"CLOSED-BOX DESIGN"
1580 R$ = "N":L$ = "N"
1590 PRINT:PRINT:PRINT D$
1600 PRINT"QTS = "QS
1610 PRINT"QTC = ";QTC
1620 PRINT"VAS = ";VAS
1630 PRINT"VB = ";VB
1640 PRINT"FS = ";FS;" HZ"
1650 PRINT"ALPHA = ";A
1660 PRINT"FC = ";FC;" HZ"
1670 PRINT"F3 = ";F3;" HZ"
1680 PRINT:INPUT"LINE PRINT OUTPUT (Y)";L$
1690 IF L$ <> "Y" THEN 1780
1700 LPRINT TAB(22)"CLOSED BOX DESIGN":LPRINT:LPRINT
1710 LPRINT"QTS = ";QS
1720 LPRINT"QTC = ";QTC
1730 LPRINT"VAS = ";VAS
1740 LPRINT"VB = ";VB
1750 LPRINT"FS = ";FS;" HZ"
1760 LPRINT"ALPHA = ";A
1770 LPRINT"FC = ";FC;" HZ"
1775 LPRINT "F3 = ";F3;" HZ"
1780 PRINT:INPUT"CLOSED BOX RESPONSE GRAPH (Y)";R$
1790 IF R$ <> "Y" THEN 1865
1800 CLS:PRINT TAB(15)"GRAPHICS CALCULATIONS IN PROGRESS"
1810 FOR F = 20 TO 200 STEP 5
1820 FH = F/FC:FQ = FH^2:MAG = FQ/(SQR((FQ-1)^2+(FH/QTC)^2))
1830 M(F) = 20*(LOG(MAG)/LOG(10))
1840 NEXT
1850 GOSUB 1900
1860 GOTO 1570
1865 R$ = "N"
1870 INPUT"ANOTHER CLOSED BOX DESIGN";R$
1880 IF R$ = "Y" THEN 1370
1890 GOTO 40
1900 CLS:RESTORE:R$ = "N"
1910 PRINT:PRINT:PRINT:PRINT:PRINT"0 DB":PRINT:PRINT:PRINT"5 DB":PRINT"PER":PRIN
T"DIV"
1920 PRINT:PRINT:PRINT:PRINT:PRINT
1930 PRINT TAB(5)"20" TAB(25)"HZ" TAB(45)"200"
1940 FOR I = 12 TO 120
1950 PSET(4*I,Y*38)
1960 NEXT
1970 FOR I = 3 TO 38
1980 PSET(43,Y*I)
1990 NEXT
2000 FOR I = 3 TO 33 STEP 5
2010 PSET(44,Y*I)
2015 PSET(45,Y*I)
2020 NEXT
2030 FOR I = 1 TO 9
2040 READ J
2050 PSET(X*((LOG(J/20)*18.3851)+12),(Y*38))
2052 PSET(X*((LOG(J/20)*18.3851)+12),(Y*37.5))
2060 NEXT
2070 FOR F = 20 TO 200 STEP 5
```

```
2080 IF M(F) <-25 THEN M(F) = -25
2090 PSET(X*((LOG(F/20)*18.3851)+12),Y*((38-M(F))-25))
2100 NEXT
2110 DATA 30,40,50,60,70,80,90,100,200
2115 DATA 20,30,40,45,50,60,80,100,120,140,160,200
2120 INPUT"LINE PRINT (Y)   ";R$
2130 IF R$ = "Y" THEN 2150
2140 RETURN
2150 LPRINT TAB(9)"-40" TAB(19)"-30" TAB(29)"-20" TAB(39)"-10" TAB(50)"0" TAB(59
)"10"
2160 LPRINT TAB(10)"+" TAB(20)"+" TAB(30)"+" TAB(40)"+" TAB(50)"+" TAB(60)"+"
2170 FOR I = 1 TO 12
2180 READ F
2190 LPRINT F TAB(9)"I";
2200 IF M(F)<(-39) THEN 2230
2210 LPRINT TAB(50+M(F))"*"
2220 GOTO 2240
2230 LPRINT""
2240 NEXT
2250 INPUT"PRESS ENTER TO CONTINUE";R$
2260 RETURN
```

# Glossary

**alignment**  A class of enclosure parameters that provides optimum performance for a woofer with a given value of Q.

**axial mode**  The enclosure resonance mode that corresponds to waves traveling the length of the enclosure and are reflected between opposite ends.

**baffle**  A board or other plane surface that holds a loudspeaker.

**bandwidth**  The range of frequencies covered by a driver or a network.

**beaming**  The tendency of a loudspeaker to concentrate the sound in a narrow path instead of spreading it.

**bi-wiring**  A method of connecting the amplifier or receiver to the speaker where a separate wire is run to each driver in the system.

**Butterworth network**  An audio filter with a flat response curve and a Q of 0.707. Second-order Butterworth crossovers give a flat power response but not a flat frequency magnitude response.

**channel**  The path an audio signal travels through a circuit during recording or playback. At least two channels are required for stereo sound.

**closed-box speaker**  A speaker system with an airtight enclosure that completely isolates the back wave from the front.

**coloration**  Any change in the character of sound that reduces naturalness such as an overemphasis of certain tones.

**compliance**  The degree to which a loudspeaker diaphragm yields elastically when a force is applied to it. The opposite of stiffness.

**cone**   The shape of the diaphragm of most woofers, full-range speakers, some midrange drivers, and some low-priced tweeters.

**constant ratio**   A ratio that remains constant in more than one application. For example, a speaker enclosure that has a length/width ratio equal to the width/depth ratio is a constant-ratio enclosure.

**crossover frequency**   The dividing frequency at which the woofer and tweeter receive equal power from the crossover network.

**crossover network**   The network that divides the audio signal to send only low frequencies to the woofer and high frequencies to the tweeter.

**damping**   The control of vibration or excursion by electrical or mechanical means.

**damping material**   Any material added to the interior of a speaker enclosure to absorb sound and reduce out-of-phase reflection to the driver diaphragm.

**decibel (dB)**   The unit of change of loudness. One dB represents the smallest change that can be heard.

**decay time**   The length of time required for a sound to decay to a level 60 dB below that of the initial sound level. As a practical measurement, it is the length of time you can hear a sound pulse.

**diaphragm**   The part of a dynamic loudspeaker attached to the voice coil that moves and produces the sound. It usually has the shape of a cone or dome.

**diffraction**   A change in the direction of a wave front that is caused by the wave moving past an obstacle.

**dispersion**   The degree that sound spreads over the listening area.

**displacement**   The volume of space occupied by a diaphragm as it goes through one cycle of movement. It can be determined by multiplying the effective piston (cone) area by the distance traveled.

**distortion**   Any error in the reproduction of sound that alters the original signal.

**dome**   A popular shape for midrange and tweeter loudspeakers. The voice coil is the same diameter as the dome, driving it from the periphery instead of from the center as in a cone. The dome shape is particularly useful for tweeters because most cone tweeters show breakup in their upper ranges. When breakup occurs, parts of the diaphragm move out of phase with other parts.

**dope**   A tacky substance added to paper cones to damp spurious vibrations that can cause breakup and rough response.

**double voice coil**   A voice coil with two windings. A subwoofer with a double voice coil can receive bass frequencies from the two stereo channels, saving the space needed by a second woofer.

**driver**   A loudspeaker unit, such as a woofer or tweeter.

**driver parameters**   The physical properties of a driver that determine its electrical and acoustical behavior.

**dynamic driver**   A loudspeaker that consists of a frame that holds a magnet and a suspended diaphragm. A voice coil, attached to the diaphragm, receives the electrical signal from the amplifier and moves in the magnetic field. The movement of the diaphragm produces sound.

**dynamic range**   The range of sound intensity a system can reproduce without compressing or distorting the signal.

**efficiency rating**   The loudspeaker parameter that shows the level of sound output when measured at a prescribed distance with a standard level of electrical energy fed into the speaker.

**electrostatic speaker**   A loudspeaker in which a thin, conductive diaphragm is suspended between two charged electrodes.

**enclosure**   The box that contains the woofer, tweeter, and any other drivers used in a speaker system.

**$f_3$**   The frequency at which the driver's response is down 3 dB from the level of its midband response, sometimes called the *cutoff frequency.*

**$f_B$**   The tuned frequency of a ported box.

**$f_C$ or $f_{CB}$**   The system resonance frequency of a driver in a closed box.

**$f_s$**   The frequency of resonance for a driver in free air.

**filter**   Any electrical circuit or mechanical device that removes or attenuates energy at certain frequencies.

**first-order network**   A crossover network that has one element in each branch of the circuit and produces a roll-off of 6 dB per octave.

**flat response**   The reproduction of sound without altering the intensity of any part of the frequency range.

**flush mounting**   Installing a driver so that its front surface is even with the front surface of the speaker board.

**fourth-order network**   A crossover network that has four elements in each branch and produces a roll-off of 24 dB per octave.

**frequency**   The number of vibrations per second. In audio, the identification of an audible tone by the number of vibrations per second.

**frequency response**   The frequency range to which a system, or any part of it, can respond. Unless a limit of variation in intensity is stated, this specification is meaningless.

**Hertz (Hz)**   A measurement of the frequency of sound vibration. One hertz is equal to one cycle per second. The hertz is named for H. R. Hertz, a German physicist.

**high-pass filter**   A filter that passes high frequencies but blocks low ones.

**image**   The mental picture of the precise location of the various instruments detectable while listening to a stereo system.

**impedance**   The combined effect of a speaker's resistance, inductance, and capacitance that opposes the current fed to it. It is measured in ohms and varies with the frequency of the signal.

**infinite baffle**   A flat surface, such as a room wall, that completely isolates the back wave of a loudspeaker from the front. Some early, large, closed-box speakers were called infinite baffles.

**infrasonic filter**   A filter designed to remove extremely low-frequency noise from the audio signal. Signals at frequencies below 20 Hz often consist chiefly of turntable rumble or other such noise and yet can overload the speaker system at ordinary listening levels.

**input**   The current fed into the loudspeaker.

**kilohertz (kHz)**   One thousand hertz.

**$L_e$**   The inductance of a driver's voice coil.

**lobing**   The tendency of a speaker system that consists of more than one driver to produce a lobed frequency response in space with in-phase reinforcement (lobes) from the various drivers occurring at some elevations and out-of-phase opposition (nulls) at points between the lobes (see Fig. 8-2).

**Linkwitz-Riley network**   A squared Butterworth network with a Q of 0.5. Second-order L-R networks have a flat frequency magnitude response but not a flat power response.

**L-pad**   A variable resistance control that puts one element of resistance in series with the load and another element in parallel. This permits the L-pad to vary the signal to a driver while maintaining a constant impedance at its input.

**low-pass filter**   A filter that passes low frequencies but blocks high ones.

**midrange**   The frequency range above bass but below treble that carries most of the identifying tones of music or speech. It is variously defined to be from 300 to 500 Hz at the lower end to 3 to 5 kHz at the top.

**midrange driver**   A loudspeaker designed specifically to reproduce the midrange frequency band. The typical range of cone midrange drivers is 300 Hz to 5 kHz, whereas dome types usually have a more limited low-frequency range.

**monitor speaker** A high-quality speaker used to assess the performance of an electronic system, broadcast signal, or recording process.

**mono** Monophonic sound. A method of recording or reproducing sound in which the signals from all directions or sources are blended into a single channel.

**oblique mode** The resonance mode of an enclosure that corresponds to waves that are reflected at oblique angles from all six walls of the enclosure.

**octave** A range of tones where the highest tone occurs at twice the frequency of the lowest tone.

**output** The sound level produced by a loudspeaker.

**passive radiator** A diaphragm without a voice coil or magnet, sometimes called a *drone cone*.

**phase distortion** A type of audible distortion caused by time delay between various parts of the signal.

**pink noise** Noise in which the amplitude is inversely proportional to frequency, giving a mellow effect.

**pipe** An acoustic device based on the organ pipe. Normally, pipes are highly resonant unless stuffed with damping material.

**polarity** The 180-degree difference in the phase of audio signals that must be observed when wiring speakers.

**ported box speaker** A speaker system with a ported enclosure that uses the back wave to reinforce the bass. It is sometimes called a *bass reflex speaker*.

**power** The measure of energy at the input or output of a device. The power rating of a speaker is usually an indication of the maximum power that can be safely fed into the speaker. The power output of the speaker depends on its efficiency.

**Q** The magnification of resonance factor of any resonant device or circuit. A driver with a high Q is more resonant than one with a low Q.

**$Q_{ES}$** The electrical Q of the driver.

**$Q_{MS}$** The mechanical Q of the driver.

**$Q_{TS}$** The total Q of the driver at $f_s$.

**reactor** An element used in a crossover network whose reactance value varies with the frequency of the signal.

**resonance** The property of a system to vibrate more at a certain frequency than at any other frequency.

**resonance frequency** The frequency at which any system vibrates naturally when excited by a stimulus. A tuning fork, for example, resonates at a specific frequency when struck.

**reverberation** Occurs when reflections of a sound within a closed space "stretch" the time the sound is audible.

**reverberation time**   See *decay time.*

**ribbon speaker**   A loudspeaker that consists of a thin, corrugated, metallic ribbon suspended in a magnetic field. The ribbon acts electrically like a low-impedance voice coil and mechanically as a diaphragm.

**roll-off**   The attenuation that occurs at the lower or upper frequency range of a driver, network, or system. The roll-off frequency is usually defined as the frequency where response is reduced by 3 dB.

**second-order network**   A crossover network that has two elements in each branch, producing a roll-off of 12 dB per octave.

**soft dome tweeter**   A tweeter that has a fabric dome as the radiating diaphragm.

**speaker system**   The arrangement of drivers, the crossover network, and the enclosure designed to perform as a whole to reproduce sound as naturally as possible (for the drivers used).

**sound pressure level (SPL)**   The intensity, or level, of sound as measured on the dB scale. A sound that has an intensity 10 times as great as another sound differs from it by 10 dB.

**stereo**   Stereophonic sound. A method of recapturing the depth and breadth of a live performance by recording and reproducing the sound in two separate channels set apart in space.

**standing wave**   A buildup of sound level at a particular frequency that is dependent upon the dimensions of a resonant room or enclosure. It occurs when the rate of energy loss equals the rate of energy input into the system. This is what you hear when you listen into a sea shell.

**subwoofer**   A loudspeaker designed to reproduce a range of very low frequencies only. The typical range of a subwoofer is 20 to 200 Hz.

**suspension**   The part of a loudspeaker that holds the diaphragm in place but allows it to move when activated.

**tangential mode**   The resonance mode of an enclosure that corresponds to waves that are reflected from the four walls of the enclosure and travel parallel to the other two walls.

**terminals**   The points where the cable from the amplifier or receiver are connected to the speaker system.

**transient response**   The ability of a speaker to respond to any sudden change in the signal without blurring the sound.

**third-order network**   A crossover network that has three elements in each branch and produces a rolloff of 18 dB per octave.

**transmission line speaker**   A speaker system installed in an enclosure that has a long tunnel behind the woofer. The tunnel is stuffed with damping material to attenuate all but the lowest frequencies from the back wave of the woofer. The low-frequency sound can be allowed to emerge from the end of the tunnel to reinforce the bass.

**tweaking** Any refinement that an audio professional or hobbyist can make to improve the naturalness of sound reproduction from a system no matter how great the effort nor how small the improvement.

**tweeter** A loudspeaker designed to reproduce high frequencies only. The typical range of small dome tweeters is from about 3 kHz upward, but some tweeters are used as low as 1 kHz.

**V$_{AS}$** The volume of air that offers the same degree of restoring force on the cone as that of the cone's suspension.

**voice coil** The cylindrical coil of wire that moves in the magnetic field of a dynamic driver. The voice coil is cemented to the diaphragm that produces the sound.

**wavelength** The length of a sound wave in air. It can be found for any frequency by dividing the speed of sound in air (1120 feet per second) by the frequency of the sound.

**white noise** Random noise equally distributed over the audible frequency range, from 20 Hz to 20 kHz. A white noise test can be approximated by using interstation hiss from an FM tuner or receiver.

**woofer** A bass loudspeaker designed to reproduce low-frequency sound only. Large woofers have a typical frequency range of 20 Hz to 1 kHz. Small woofers often have a range of 50 Hz to 3 kHz.

# Index

# New Loudspeaker Design Program Disk
# Version 1.1

This disk includes the original Loudspeaker Design Program plus three new sections. The new additions permit you to quickly compute port dimensions, design double-chamber reflex enclosures, and obtain up to five sets of recommended box dimensions ratios for any desired cubic volume.

    If the idea of typing the Loudspeaker Design Program in the Appendix seems too laborious, you can order ready-to-run disks. The disks can be used with the IBM PC/XT/AT and their compatibles or the PS/2. You'll need 256K, GWBASIC or BASICA, and DOS 2.1 or later; the optional graphics require a CGA (or better) color video adapter. The cost is $19.95 for either a 5¼" or a 3½" floppy plus $2.50 ($5.00 outside the U.S.) for shipping and handling per order.

---

YES, I'm interested. Please send me:

_____ copies 5¼" Loudspeaker Design Program disk(s), $19.95 each . . . . . . . $ _____

_____ copies 3½" Loudspeaker Design Program disks(s), $19.95 each . . . . . . $ _____

_____ TAB Books catalog (free with purchase; otherwise send $1.00
in check or money order and receive coupon worth $1.00 off your next
purchase) . . . . . . . . . . . . . . . . . . . . . . . . . . . . . . . . . . . . . . . . . . . . . . . $ _____

Shipping & Handling: $2.50 per disk in U.S.
($5.00 per disk outside U.S.)   $ _____

Please add applicable state and local sales tax.   $ _____

TOTAL   $ _____

☐ Check or money order enclosed made payable to TAB Books

Charge my  ☐ VISA  ☐ MasterCard  ☐ American Express

Acct No. _____ Exp. Date _____

Signature _____

Name _____

Address _____

City _____ State _____ Zip _____

## TOLL-FREE ORDERING: 1-800-822-8158
(in PA, AK, and Canada call 1-717-794-2191)
or write to TAB Books, Blue Ridge Summit, PA 17294-0840

Prices subject to change. Orders outside the U.S. must be paid in international money order in U.S. dollars.

TAB-3274